MW01325250

THE FOUNDATIONS OF MATH

GEOMETRY

EDITED BY NICHOLAS FAULKNER AND
WILLIAM L. HOSCH

IN ASSOCIATION WITH

Published in 2018 by Britannica Educational Publishing (a trademark of Encyclopædia Britannica, Inc.) in association with The Rosen Publishing Group, Inc.
29 East 21st Street, New York, NY 10010

Distributed exclusively by Rosen Publishing.
To see additional Britannica Educational Publishing titles, go to rosenpublishing.com.

Britannica Educational Publishing
J.E. Luebering: Executive Director, Core Editorial
Andrea R. Field: Managing Editor, Compton's by Britannica

Rosen Publishing
Nicholas Faulkner: Editor
Nelson Sá: Art Director
Brian Garvey: Series Designer
Tahara Anderson: Book Layout
Cindy Reiman: Photography Manager

Library of Congress Cataloging-in-Publication Data

Names: Faulkner, Nicholas, editor. | Hosch, William L., editor.
Title: Geometry / edited by Nicholas Faulkner and William L. Hosch.
Description: New York : Britannica Educational Publishing, in Association with Rosen Educational Services, 2018. | Series: The foundations of math | Audience: Grades 9-12. | Includes bibliographical references and index.
Identifiers: LCCN 2017018569 | ISBN 9781680487763 (library bound : alk. paper)
Subjects: LCSH: Geometry--Juvenile literature. | Geometry--History--Juvenile literature. | Geometry--Biography--Juvenile literature.
Classification: LCC QA445.5 .G445 2018 | DDC 516--dc23
LC record available at https://lccn.loc.gov/2017018569

Manufactured in Malaysia

Photo credits: Cover White Lace Photo/Shutterstock.com; pp. 32, 43, 53, 65, 67, 75, 102, 107, 112, 114, 126 Adapted from information from Encyclopaedia Britannica, Inc.; p. 219 Getty Images; p. 254 © Photos.com/Thinkstock; p. 271 Courtesy of the Musée des Beaux-Arts, Beaune, France; pp. 282-283 Scanpix/AP Images; p. 333 Archiv für Kunst und Geschichte. All other illustrations, diagrams, graphs, maps, formulas, and equations Encyclopædia Britannica, Inc.

CONTENTS

INTRODUCTION

Geometry is the branch of mathematics concerned with the shape of individual objects, spatial relationships among various objects, and the properties of surrounding space. It is one of the oldest branches of mathematics, having arisen in response to such practical problems as those found in surveying, and its name is derived from Greek words meaning "Earth measurement." Eventually it was realized that geometry need not be limited to the study of flat surfaces (plane geometry) and rigid three-dimensional objects (solid geometry) but that even the most abstract thoughts and images might be represented and developed in geometric terms.

In several ancient cultures there developed a form of geometry suited to the relationships between lengths, areas, and volumes of physical objects. This geometry was codified in the Greek mathematician Euclid's *Elements* about 300 BCE on the basis of 10 axioms, or postulates, from which several hundred theorems were proved by deductive logic. The *Elements* epitomized the axiomatic-deductive method for many centuries.

Analytic geometry was initiated by the French mathematician René Descartes (1596–1650), who introduced rectangular coordinates to locate points and to enable lines and curves to be represented with algebraic equations. Algebraic geometry is a modern extension of the subject to multidimensional and non-Euclidean spaces.

Projective geometry originated with the French mathematician Girard Desargues (1591–1661) to deal with those properties of geometric figures that are not altered by projecting their image, or "shadow," onto another surface.

The German mathematician Carl Friedrich Gauss (1777–1855), in connection with practical problems of surveying and geodesy, initiated the field of differential geometry. Using differential calculus, he characterized the intrinsic properties of curves and surfaces. For instance, he showed that the intrinsic curvature of a cylinder is the same as that of a plane, as can be seen by cutting a cylinder along its axis and flattening, but not the same as that of a sphere, which cannot be flattened without distortion.

Beginning in the 19th century, various mathematicians substituted alternatives to Euclid's parallel postulate, which, in its modern form, reads, "given a line and a point not on the line, it is possible to draw exactly one line through the given point parallel to the line." They hoped to show that the alternatives were logically impossible. Instead, they discovered that consistent non-Euclidean geometries exist.

Topology, the youngest and most sophisticated branch of geometry, focuses on the properties of geometric objects that remain unchanged upon continuous deformation—shrinking, stretching, and folding, but not tearing. The continuous development of topology dates from 1911, when the Dutch mathematician L.E.J. Brouwer (1881–1966) introduced methods generally applicable to the topic.

Geometry was thoroughly organized in about 300 BCE, when Euclid gathered what was known at the time, added original work of his own, and arranged 465 propositions into 13 books, collectively called *Elements*. The books covered not only plane and solid geometry but also much of what is now known as algebra, trigonometry, and advanced arithmetic.

Down through the ages, the propositions have been rearranged, and many of the proofs are different, but the basic idea presented in the *Elements* has not changed. In the work facts are not just cataloged but are developed in an orderly way, starting with statements (definitions, common notions, and postulates) that seem perfectly self-evident, with each successive theorem proved by using only previously shown facts. This mode of reasoning, known as the axiomatic method, has profoundly influenced epistemology (the study of the nature, origin, and limits of human knowledge) and education.

Even in 300 BCE, geometry was recognized to be not just for mathematicians. Anyone can benefit from the basic teachings of geometry, which are how to follow lines of reasoning, how to say precisely what is intended, and especially how to prove basic concepts by following these lines of reasoning.

Geometry in ancient times was recognized as part of everyone's education. Early Greek philosophers asked that no one come to their schools who had not learned the *Elements* of Euclid. There were, and still are, many who resisted this kind of education. It is said that Ptolemy I asked Euclid for an easier way to learn the material. Euclid told him there was no "royal road" to geometry. The same message applies to readers of this volume. They will not learn what geometry is all about. What they will learn is the basic shapes of some of the figures dealt with in geometry and a few facts about them. It takes a geometry course, with textbook and teacher, to show the complete and orderly arrangement of the facts and how each is proved.

Euclid's fifth postulate asserts that, given a line and a point not on the line, there exists a unique line through the point and parallel to the given line. This postulate never seemed completely obvious, and mathematicians strove for centuries to find a proof for it based on Euclid's other, more obviously true, postulates. With no such direct proof forthcoming, some mathematicians began by assuming a different fifth postulate—either that there are infinitely many parallel lines (hyperbolic geometry) or that there are no parallel lines (elliptic geometry)—in the hopes of discovering a logical contradiction which would thereby indirectly prove Euclid's fifth. A particularly famous example of this was a flawed proof in 1733 by the Italian Girolamo Saccheri, based on the quadrilateral figure of the Persian Omar Khayyam (from about the year 1000).

There was no further progress until the Russian Nikolay Lobachevsky published the first paper on hyperbolic geometry in 1829. Although Lobachevsky continued his research in "imaginary geometry" for more than a decade, his work was not widely known or respected. It had little impact before the German Bernhard Riemann developed an axiomatic system for elliptic geometry in the 1850s. Suddenly there were three incompatible geometries and a loss of certainty in geometry as the realm of indisputable knowledge. In 1871 the German Felix Klein compounded the problem by showing that all of these alternate geometries were internally consistent, leaving open the question of which one corresponds with reality. Near the beginning of the 20th century, Albert Einstein incorporated Riemann's work in his mathematical description of his theory of relativity (involving curved, or Riemannian, space).

The roots of elliptic geometry go back to antiquity in the form of spherical geometry. In spherical geometry everything resides on the surface of a sphere, making spherical geometry central for cartography and astronomy. "Lines" are defined as the great circles—circles whose centers coincide with the sphere's center, dividing it into hemispheres. Great circles can form angles and triangles and other polygons. They can, in fact, do anything that lines do on a plane except be parallel. Any two great circles meet each other at two diametrically opposite points. Small circles (for example, the lines of latitude on Earth above and below the Equator) can be parallel, but they do not have the other properties of straight lines.

One of the better-known facts of Euclidean geometry is that the angles of a triangle add up to one straight angle, or 180°. This may appear to have nothing to do with parallel lines, but the relationship cannot be proved without Euclid's parallel postulate. A special definition of angles must be used for spherical geometry because the directions of lines change (as viewed from outside the surface). The angles of a spherical triangle always add up to more than 180°. The larger the triangle relative to the sphere, the greater the amount by which the sum exceeds 180°.

"Saddle" geometry, whose development is generally attributed to the greatest mathematician of the 19th century, the German Carl Friedrich Gauss, is based on a surface that curves in two directions at once, like a saddle or certain mountain passes. Such a surface cannot be extended indefinitely, like a plane, nor can it meet itself in a shape as tidy as a sphere. The angles of a triangle drawn on the surface of a saddle-like mountain pass add up to less than 180°. The figure shows a natural rock formation that is much like a triangle in saddle geometry. While spherical geometry has no parallels, in saddle geometry many lines can be drawn through the same point, all parallel to the same line. This work of Gauss, published after his death in 1855, led many mathematicians to take non-Euclidean geometry seriously.

In 1975 Polish mathematician Benoit Mandelbrot introduced fractal geometry as a way to describe irregularly shaped objects or natural phenomena—such as coastlines, snowflakes, and tree branches—that could not be described by Euclidean geometry. Mandelbrot coined the word "fractal" to signify certain complex geometric shapes. The word is derived from the Latin *fractus*, meaning "fragmented" or "broken" and refers to the fact that these objects are self-similar—that is, their component parts resemble the whole. He stated that natural forms have the tendency to repeat themselves on an ever smaller scale, so that if each component is magnified it will look basically like the object as a whole. This geometry has been applied to the fields of physiology, chemistry, and mechanics.

Taking a course in geometry is beneficial for all students, who will find that learning to reason and prove convincingly is necessary for every profession. It is true that not everyone must prove things, but everyone is exposed to proof. Politicians, advertisers, and many other people try to offer convincing arguments. Anyone who cannot tell a good proof from a bad one may easily be persuaded in the wrong direction. Geometry provides a simplified universe, where points and lines obey believable rules and where conclusions are easily verified. By first studying how to reason in this simplified universe, people can eventually, through practice and experience, learn how to reason in a complicated world.

Chapter 1

HISTORY OF GEOMETRY

The earliest known unambiguous examples of written records—dating from Egypt and Mesopotamia about 3100 BCE—demonstrate that ancient peoples had already begun to devise mathematical rules and techniques useful for surveying land areas, constructing buildings, and measuring storage containers. Beginning about the 6th century BCE, the Greeks gathered and extended this practical knowledge and from it generalized the abstract subject now known as geometry, from the combination of the Greek words *geo* ("Earth") and *metron* ("measure") for the measurement of the Earth.

ANCIENT GEOMETRY: PRACTICAL AND EMPIRICAL

The origin of geometry lies in the concerns of everyday life. The traditional account, preserved in Herodotus's *History* (5th century BCE), credits the Egyptians with inventing surveying in order to reestablish property values after the annual flood of the Nile. Similarly,

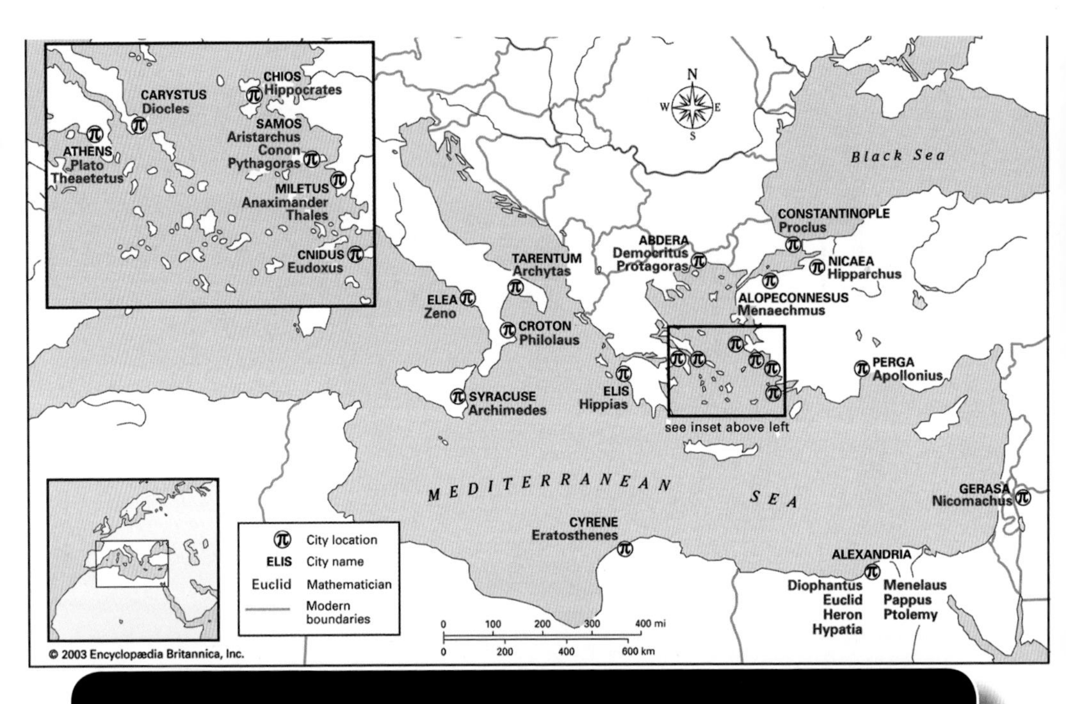

This map spans a millennium of prominent Greco-Roman mathematicians.

eagerness to know the volumes of solid figures derived from the need to evaluate tribute, store oil and grain, and build dams and pyramids. Even the three abstruse geometrical problems of ancient times—to double a cube, trisect an angle, and square a circle, all of which will be discussed later—probably arose from practical matters, from religious ritual, timekeeping, and construction, respectively, in pre-Greek societies of the Mediterranean. and the main subject of later Greek geometry, the theory of conic sections, owed its general importance, and perhaps also its origin, to its application to optics and astronomy.

While many ancient individuals, known and unknown, contributed to the subject, none equaled the impact of Euclid and his *Elements*, a book on geometry now 2,300 years old and the object of as much painful and painstaking study as the Bible. Much less is known about Euclid, however, than about Moses. In fact, the only thing known with a fair degree of confidence is that Euclid taught at the Library of Alexandria during the reign of Ptolemy I (323– 285/283 BCE). Euclid wrote not only on geometry but also on astronomy and optics and perhaps also on mechanics and music. Only the *Elements*, which was extensively copied and translated, has survived intact.

Euclid's *Elements* was so complete and clearly written that it literally obliterated the work of his predecessors. What is known about Greek geometry before him comes primarily from bits quoted by Plato and Aristotle and by later mathematicians and commentators. Among other precious items they preserved are some results and the general approach of Pythagoras (c. 580–c. 500 BCE) and his followers. The Pythagoreans convinced themselves that all things are, or owe their relationships to, numbers. The doctrine gave mathematics supreme importance in the investigation and understanding of the world. Plato developed a similar view, and philosophers influenced by Pythagoras or Plato often wrote ecstatically about geometry as the key to the interpretation of the universe. Thus ancient geometry gained an association with the sublime to complement its earthy origins and its reputation as the exemplar of precise reasoning.

FINDING THE RIGHT ANGLE

Ancient builders and surveyors needed to be able to construct right angles in the field on demand. The method employed by the Egyptians earned them the name "rope pullers" in Greece, apparently because they employed a rope for laying out their construction guidelines. One way that they could have employed a rope to construct right triangles was to mark a looped rope with knots so that, when held at the knots and pulled tight, the rope must form a right triangle. The simplest way to perform the trick is to take a rope that is 12 units long, make a knot 3 units from one end and another 5 units from the other end, and then knot the ends together to form a loop. However, the Egyptian scribes have not left us instructions about these procedures, much less any hint that they knew how to generalize them to obtain the Pythagorean theorem: the square on the line opposite the right angle equals the sum of the squares on the other two sides. Similarly, the Vedic scriptures of ancient India contain sections called sulvasutras, or "rules of the rope," for the exact positioning of sacrificial altars. The required right angles were made by ropes marked to give the triads (3, 4, 5) and (5, 12, 13).

In Babylonian clay tablets (c. 1700–1500 BCE) modern historians have discovered problems whose solutions indicate that the Pythagorean theorem and some special triads were known more than 1,000 years before Euclid. A right triangle made at random, however, is very unlikely to have all its sides measurable by the same unit—that is, every side a whole-number multiple of some common unit of measurement. This fact, which came as a shock when discovered by the Pythagoreans, gave rise to the concept and theory of incommensurability.

Locating the Inaccessible

By ancient tradition, Thales of Miletus, who lived before Pythagoras in the 6th century BCE, invented a way to measure inaccessible heights, such as the Egyptian pyramids. Although none of his writings survives, Thales may well have known about a Babylonian observation that for similar triangles (triangles having the same shape but not necessarily the same size) the length of each corresponding side is increased (or decreased) by the same multiple. The ancient Chinese arrived at measures of inaccessible heights and distances by another route, using "complementary" rectangles, which can be shown to give results equivalent to those of the Greek method involving triangles.

Estimating the Wealth

A Babylonian cuneiform tablet written some 3,500 years ago treats problems about dams, wells, water clocks, and excavations. It also has an exercise on circular enclosures

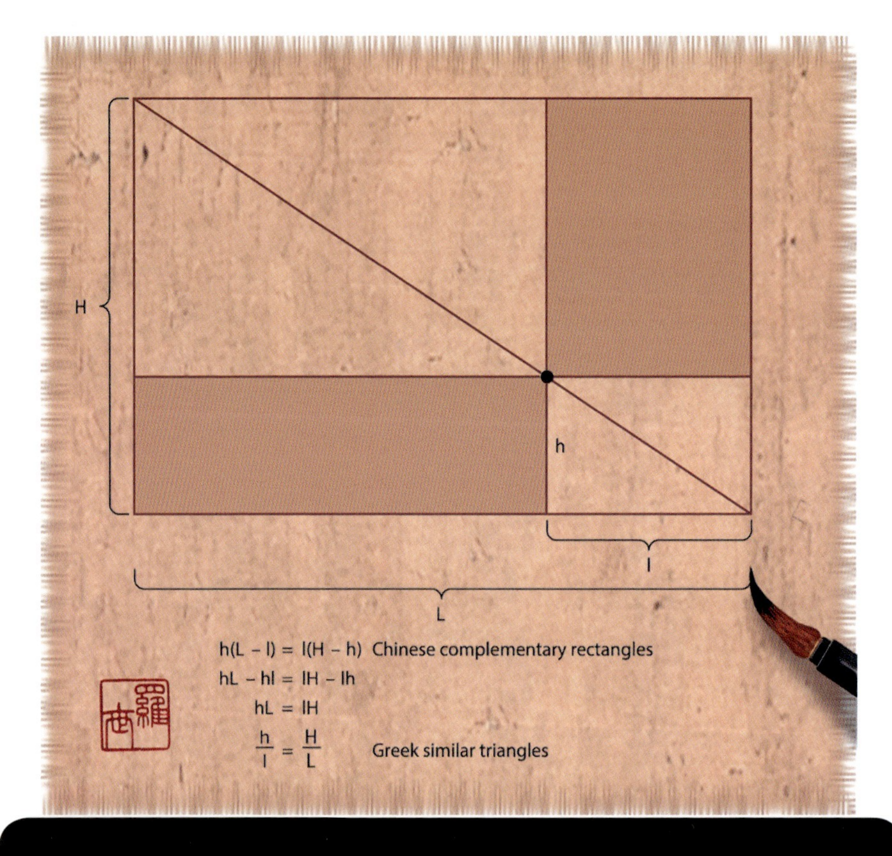

The equivalence of the Chinese complementary rectangles theorem and the Greek similar triangles theorem.

with an implied value of $\pi = 3$. The contractor for King Solomon's swimming pool, who made a pond 10 cubits across and 30 cubits around (1 Kings 7:23), used the same value. However, the Hebrews should have taken their π from the Egyptians before crossing the Red Sea, for the Rhind papyrus (c. 2000 BCE; our principal source for ancient Egyptian mathematics) implies $\pi = 3.1605$.

Knowledge of the area of a circle was of practical value to the officials who kept track of the pharaoh's tribute as well as to the builders of altars and swimming pools. Ahmes, the scribe who copied and annotated the Rhind papyrus (c. 1650 BCE), has much to say about cylindrical granaries and pyramids, whole and truncated. He could calculate their volumes, and, as appears from his taking the Egyptian seked, the horizontal distance associated with a vertical rise of one cubit, as the defining quantity for the pyramid's slope, he knew something about similar triangles.

Ancient Geometry: Abstract and Applied

In addition to proving mathematical theorems, ancient mathematicians constructed various geometrical objects. Euclid arbitrarily restricted the tools of construction to a straightedge (an unmarked ruler) and a compass. The restriction made three problems of particular interest (to double a cube, to trisect an arbitrary angle, and to square a circle) very difficult—in fact, impossible. Various methods of construction using other means were devised in the classical period, and efforts, always unsuccessful, using a straightedge and a compass persisted for the next 2,000 years. In 1837 the French mathematician Pierre Laurent Wantzel proved that doubling the cube and trisecting the angle are impossible, and in 1880 the German mathematician Ferdinand von Lindemann showed that squaring the circle is impossible, as a consequence of his proof that π is a transcendental number.

DOUBLING THE CUBE

The Vedic scriptures made the cube the most advisable form of altar for anyone who wanted to supplicate in the same place twice. The rules of ritual required that the altar for the second plea have the same shape but twice the volume of the first. If the sides of the original and derived altars are a and b, respectively, then $b^3 = 2a^3$. The problem came to the Greeks together with its ceremonial content. An oracle disclosed that the citizens of Delos could free themselves of a plague merely by replacing an existing altar by one twice its size. The Delians applied to Plato. He replied that the oracle did not mean that the gods wanted a larger altar but that they had intended "to shame the Greeks for their neglect of mathematics and their contempt for geometry." With this blend of Vedic practice, Greek myth, and academic manipulation, the problem of the duplication of the cube took a leading place in the formation of Greek geometry.

Hippocrates of Chios, who wrote an early *Elements* about 450 BCE, took the first steps in cracking the altar problem. He reduced the duplication to finding two mean proportionals between 1 and 2, that is, to finding lines x and y in the ratio $1:x = x:y = y:2$. After the intervention of the Delian oracle, several geometers around

Plato's Academy found complicated ways of generating mean proportionals.

A few generations later, Eratosthenes of Cyrene (c. 276–c. 194 BCE) devised a simple instrument with moving parts that could produce approximate mean proportionals.

Trisecting the Angle

The Egyptians told time at night by the rising of 12 asterisms (constellations), each requiring on average two hours to rise. In order to obtain more convenient intervals, the Egyptians subdivided each of their asterisms into three parts, or decans. That presented the problem of trisection. It is not known whether the second celebrated problem of archaic Greek geometry, the trisection of any given angle, arose from the difficulty of the decan, but it is likely that it came from some problem in angular measure.

Several geometers of Plato's time tried their hands at trisection. Although no one succeeded in finding a solution with a straightedge and a compass, they did succeed with a mechanical device and by a trick. The mechanical device, perhaps never built, creates what the ancient geometers called a quadratrix. Invented by a geometer known as Hippias of Elis (fl. 5th century BCE), the quadratrix is a curve traced by the point of intersection between two moving lines, one rotating uniformly through a right angle, the other gliding uniformly parallel to itself.

The trick for trisection is an application of what the Greeks called neusis, a maneuvering of a measured length into a special position to complete a geometrical figure. A late version of its use, ascribed to Archimedes (c. 287–212/211 BCE), exemplifies the method of angle trisection.

SQUARING THE CIRCLE

The pre-Euclidean Greek geometers transformed the practical problem of determining the area of a circle into a tool of discovery. Three approaches can be distinguished: Hippocrates' dodge of substituting one problem for another; the application of a mechanical instrument, as in Hippias' device for trisecting the angle; and the technique that proved the most fruitful, the closer and closer approximation to an unknown magnitude difficult to study (e.g., the area of a circle) by a series of known magnitudes easier to study (e.g., areas of polygons)—a technique known in modern times as the "method of exhaustion" and attributed by its greatest practitioner, Archimedes, to Plato's student Eudoxus of Cnidus (c. 408–c. 355 BCE).

While not able to square the circle, Hippocrates did demonstrate the quadratures of lunes. That is, he showed that the area between two intersecting circular arcs could be expressed exactly as a rectilinear area and so raised the expectation that the circle itself could be treated similarly. A contemporary of Hippias' discovered that the quadratrix could be used to almost rectify circles. These were the substitution and mechanical approaches.

The method of exhaustion as developed by Eudoxus approximates a curve or surface by using polygons with calculable perimeters and areas. As the number of sides of a regular polygon inscribed in a circle increases indefinitely, its perimeter and area "exhaust," or take up, the circumference and area of the circle to within any assignable error of length or area, however small. In Archimedes' usage, the method of exhaustion produced upper and lower bounds for the value of π, the ratio of any circle's circumference to its diameter. This he accomplished by inscribing a polygon within a circle, and circumscribing a polygon around it as well, thereby bounding the circle's circumference between the polygons' calculable perimeters. He used polygons with 96 sides and thus bound π between $3^{10}/_{71}$ and $3^{1}/_{7}$.

IDEALIZATION AND PROOF

The last great Platonist and Euclidean commentator of antiquity, Proclus (c. 410–485 CE), attributed to the inexhaustible Thales the discovery of the far-from-obvious proposition that even apparently obvious propositions need proof. Proclus referred especially to the theorem, known in the Middle Ages as the Bridge of Asses, that in an isosceles triangle the angles opposite the equal sides are equal. The theorem may have earned its nickname from the Euclidean figure or from the commonsense notion that only an ass would require proof of so obvious a statement.

The ancient Greek geometers soon followed Thales over the Bridge of Asses. In the 5th century BCE the philosopher-mathematician Democritus (c. 460–c. 370 BCE) declared that his geometry excelled all the knowledge of the Egyptian rope pullers because he could prove what he claimed. By the time of Plato, geometers customarily proved their propositions. Their compulsion and the multiplication of theorems it produced fit perfectly with the endless questioning of Socrates and the uncompromising logic of Aristotle. Perhaps the origin, and certainly the exercise, of the peculiarly Greek method of mathematical proof should be sought in the same social setting that gave rise to the practice of philosophy—that is, the Greek polis. There citizens learned the skills of a governing class, and the wealthier among them enjoyed the leisure to engage their minds as they pleased, however useless the result, while slaves attended to the necessities of life. Greek society could support the transformation of geometry from a practical art to a deductive science. Despite its rigour, however, Greek geometry does not satisfy the demands of the modern systematist. Euclid himself sometimes appeals to inferences drawn from an intuitive grasp of concepts such as point and line or inside and outside, uses superposition, and so on. It took more than 2,000 years to purge the *Elements* of what pure deductivists deemed imperfections.

The Euclidean Synthesis

Euclid, in keeping with the self-conscious logic of Aristotle, began the first of his 13 books of the *Elements* with sets of definitions ("a line is breadthless length"), common notions ("the whole is greater than the part"), and axioms, or postulates ("all right angles are equal"). Of this preliminary matter, the fifth and last postulate, which states a sufficient condition that two straight lines meet if sufficiently extended, has received by far the greatest attention. In effect it defines parallelism. Many later geometers tried to prove the fifth postulate using other parts of the *Elements*. Euclid saw farther, for coherent geometries (known as non-Euclidean geometries) can be produced by replacing the fifth postulate with other postulates that contradict Euclid's choice.

The first six books contain most of what Euclid delivers about plane geometry. Book I presents many propositions doubtless discovered by his predecessors, from Thales' equality of the angles opposite the equal sides of an isosceles triangle to the Pythagorean theorem, with which the book effectively ends.

Book VI applies the theory of proportion from Book V to similar figures and presents the geometrical solution to quadratic equations. As usual, some of it is older than Euclid. Books VII–X, which concern various sorts of numbers, especially primes, and various sorts of ratios, are seldom studied now, despite the importance of the masterful Book X, with its elaborate classification of incommensurable magnitudes, to the later development of Greek geometry.

Books XI–XIII deal with solids: XI contains theorems about the intersection of planes and of lines and planes and theorems about the volumes of parallelepipeds (solids with parallel parallelograms as opposite faces); XII applies the method of exhaustion introduced by Eudoxus to the volumes of solid figures, including the sphere; XIII, a three-dimensional analogue to Book IV, describes the Platonic solids. Among the jewels in Book XII is a proof of the recipe used by the Egyptians for the volume of a pyramid.

GNOMONICS AND THE CONE

During its daily course above the horizon, the Sun appears to describe a circular arc. Supplying in his mind's eye the missing portion of the daily circle, the Greek astronomer could imagine that his real eye was at the apex of a cone, the surface of which was defined by the Sun's rays at different times of the day and the base of which was defined by the Sun's apparent diurnal course. Our astronomer, using the pointer of a sundial, known as a gnomon, as his eye, would generate a second, shadow cone spreading downward. The intersection of this second cone with a horizontal surface, such as the face of a sundial, would give the trace of the Sun's image (or shadow) during the day as a plane section of a cone. (The possible intersections of a plane with a cone, known as the conic sections, are the circle, ellipse, point, straight line, parabola, and hyperbola.)

However, the doxographers ascribe the discovery of conic sections to a student of Eudoxus's, Menaechmus (mid-4th century BCE), who used them to solve the problem of duplicating the cube. His restricted approach to conics—he worked with only right circular cones and made his sections at right angles to one of the straight lines composing their surfaces—was standard down to Archimedes' era. Euclid adopted Menaechmus's approach in his lost book on conics, and Archimedes followed suit. Doubtless, however, both knew that all the conics can be obtained from the same right cone by allowing the section at any angle.

The reason that Euclid's treatise on conics perished is that Apollonius of Perga (c. 262–c. 190 BCE) did to it what Euclid had done to the geometry of Plato's time. Apollonius reproduced known results much more generally and discovered many new properties of the figures. He first proved that all conics are sections of any circular cone, right or oblique. Apollonius introduced the terms ellipse, hyperbola, and parabola for curves produced by intersecting a circular cone with a plane at an angle less than, greater than, and equal to, respectively, the opening angle of the cone.

ASTRONOMY AND TRIGONOMETRY

In an inspired use of their geometry, the Greeks did what no earlier people seem to have done: they geometrized the heavens by supposing that the Sun, Moon, and planets move around a stationary Earth on a rotating circle or set of circles, and they calculated the speed of rotation of these supposititious circles from observed motions. Thus they assigned to the Sun a circle eccentric to the Earth to account for the unequal lengths of the seasons.

Ptolemy (fl. 127–145 CE in Alexandria, Egypt) worked out complete sets of circles for all the planets. In order to account for phenomena arising from the Earth's motion around the Sun, the Ptolemaic system included a secondary circle known as an epicycle, whose centre moved along the path of the primary orbital circle, known as the deferent. Ptolemy's Great Compilation, or *Almagest* after its Arabic translation, was to astronomy what Euclid's *Elements* was to geometry. Contrary to the *Elements*, however, the *Almagest* deploys geometry for the purpose of calculation. Among the items Ptolemy calculated was a table of chords, which correspond to the trigonometric sine function later introduced by Indian and Islamic mathematicians. The table of chords assisted the calculation of distances from angular measurements as a modern astronomer might do with the law of sines.

The application of geometry to astronomy reframed the perennial Greek pursuit of the nature of truth. If a mathematical description fit the facts, as did Ptolemy's explanation of the unequal lengths of the seasons by the eccentricity of the Sun's orbit, should the description be taken as true of nature? The answer, with increasing emphasis, was "no." Astronomers remarked that the eccentric orbit representing the Sun's annual motion could be replaced by a pair of circles, a deferent centred on the Earth and an epicycle the centre of which moved along the circumference of the deferent. That gave two observationally equivalent solar theories based on two quite different mechanisms. Geometry was too prolific of alternatives to disclose the true principles of nature. The Greeks, who had raised a sublime science from a pile of practical recipes, discovered that in reversing the process, in reapplying their mathematics to the world, they had no securer claims to truth than the Egyptian rope pullers.

Ancient Geometry: Cosmological and Metaphysical

Pythagorean Numbers and Platonic Solids

The Pythagoreans used geometrical figures to illustrate their slogan that all is number—thus their "triangular numbers" ($^{n(n-1)}/_2$), "square numbers" (n^2), and "altar numbers" (n^3). This principle found a sophisticated application in Plato's creation story, the *Timaeus*,

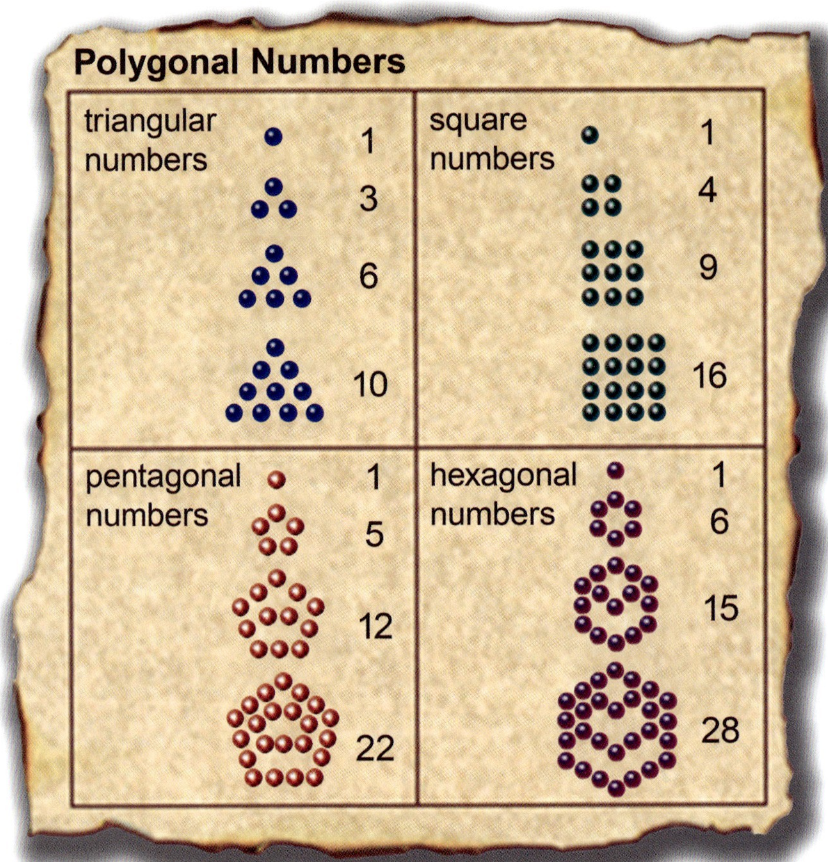

Polygonal Numbers

triangular numbers	square numbers
1, 3, 6, 10	1, 4, 9, 16
pentagonal numbers	**hexagonal numbers**
1, 5, 12, 22	1, 6, 15, 28

The ancient Greeks generally thought of numbers in concrete terms, particularly as measurements and geometric dimensions.

which presents the smallest particles, or "elements," of matter as regular geometrical figures. Since the ancients recognized four or five elements at most, Plato sought a small set of uniquely defined geometrical objects to serve as elementary constituents. He found them in the only three-dimensional structures whose faces are equal regular polygons that meet one another at equal solid angles: the tetrahedron, or pyramid (with 4 triangular faces); the cube (with 6 square faces); the octahedron (with 8 equilateral triangular faces); the dodecahedron (with 12 pentagonal faces); and the icosahedron (with 20 equilateral triangular faces).

The cosmology of the *Timaeus* had a consequence of the first importance for the development of mathematical astronomy. It guided Johannes Kepler (1571–1630) to his discovery of the laws of planetary motion. Kepler deployed the five regular Platonic solids not as indicators of the nature and number of the elements but as a model of the structure of the heavens. In 1596 he published *Prodromus Dissertationum Mathematicarum Continens Mysterium Cosmographicum* ("Cosmographic Mystery"), in which each of the known six planets revolved around the Sun on spheres separated by the five Platonic solids. Although Tycho Brahe (1546–1601), the world's greatest observational astronomer before the invention of the telescope, rejected the Copernican model of the solar system, he invited Kepler to assist him at his new observatory outside of Prague. In trying to resolve discrepancies between his original theory and Brahe's observations, Kepler made the capital discovery that the planets move in ellipses around the Sun as a focus.

MEASURING THE EARTH AND HEAVENS

Geometry offered Greek cosmologists not only a way to speculate about the structure of the universe but also the means to measure it. South of Alexandria and roughly on the same meridian of longitude is the village of Syene (modern Aswān), where the Sun stands directly overhead at noon on a midsummer day. At the same moment at Alexandria, the Sun's rays make an angle α with the tip of a vertical rod. Since the Sun's rays fall almost parallel on the Earth, the angle sub-

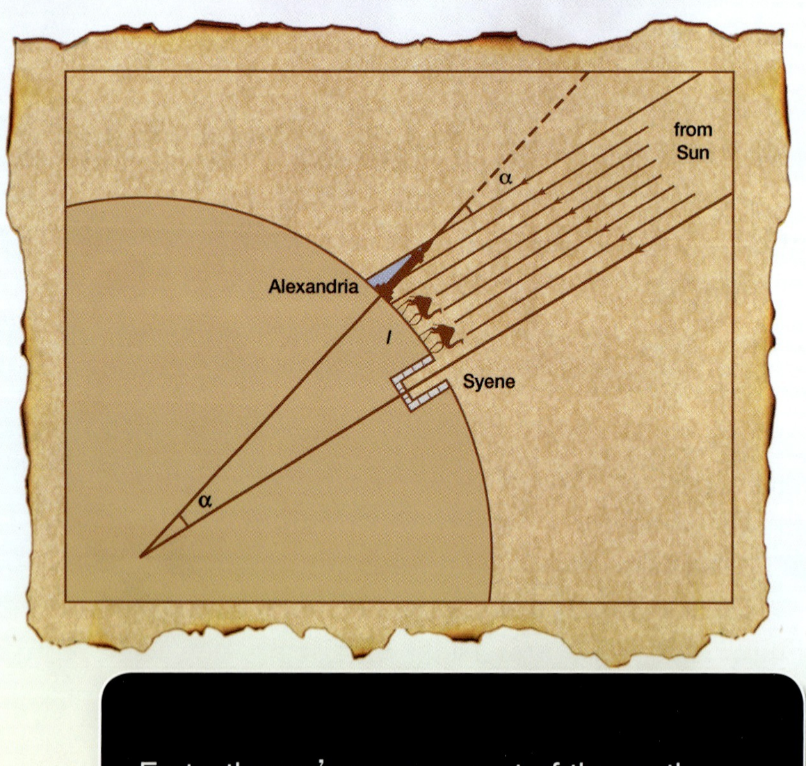

Eratosthenes' measurement of the earth.

tended by the arc *l* (representing the distance between Alexandria and Syene) at the centre of the Earth also equals α; thus the ratio of the Earth's circumference, *C*, to the distance, *l*, must equal the ratio of 360° to the angle α—in symbols, $C:l = 360°:α$. Eratosthenes made the measurements, obtaining a value of about 5,000 stadia for *l*, which gave a value for the Earth's circumference of about 400,000 stadia. Because the accepted length of the Greek stadium varied locally, we cannot accurately determine Eratosthenes' margin of error. However, if we credit the ancient historian Plutarch's guess at Eratosthenes' unit of length, we obtain a value for the Earth's circumference of about 46,250 km (28,738 miles)—remarkably close to the modern value (about 15 percent too large), considering the difficulty in accurately measuring *l* and α.

Aristarchus of Samos (c. 310–230 BCE) has garnered the credit for extending the grip of a number as far as the Sun. Using the Moon as a ruler and noting that the apparent sizes of the Sun and the Moon are about equal, he calculated values for his treatise "On the Sizes and Distances of the Sun and Moon." The great difficulty of making the observations resulted in an underestimation of the solar distance about 20-fold—he obtained a solar distance, σ, roughly 1,200 times the Earth's radius, *r*. Possibly Aristarchus' inquiry into the relative sizes of the Sun, Moon, and Earth led him to propound the first heliocentric ("Sun-centred") model of the universe.

Aristarchus' value for the solar distance was confirmed by an astonishing coincidence. Ptolemy equated the maximum distance of the Moon in its eccentric orbit with the closest approach of Mercury riding on its epicycle; the farthest distance of Mercury with the closest of Venus; and the farthest of Venus with the closest of the Sun. Thus, he could compute the solar distance in terms of the lunar distance and thence the terrestrial radius. His answer agreed with that

(Continued on the next page)

(Continued from the previous page)

of Aristarchus. The Ptolemaic conception of the order and machinery of the planets, the most powerful application of Greek geometry to the physical world, thus corroborated the result of direct measurement and established the dimensions of the cosmos for over 1,000 years. As the ancient philosophers said, there is no truth in astronomy.

The Post-Classical Period

Passage Through Islam

Two centuries after they broke out of their desert around Mecca, the followers of Muhammad occupied the lands from Persia to Spain and settled down to master the arts and sciences of the peoples they had conquered. They admired especially the works of the Greek mathematicians and physicians and the philosophy of Aristotle. By the late 9th century they were already able to add to the geometry of Euclid, Archimedes, and Apollonius. In the 10th century they went beyond Ptolemy. Stimulated by the problem of finding the effective orientation for prayer (the qiblah, or direction from the place of worship to Mecca), Islamic geometers and astronomers developed the stereographic projection (invented to project the celestial sphere onto a two-dimensional map or instrument) as well as plane and spherical trigonometry. Here they incorporated elements derived from India as well as from Greece. Their achievements in geometry and geometrical astronomy materialized in instruments

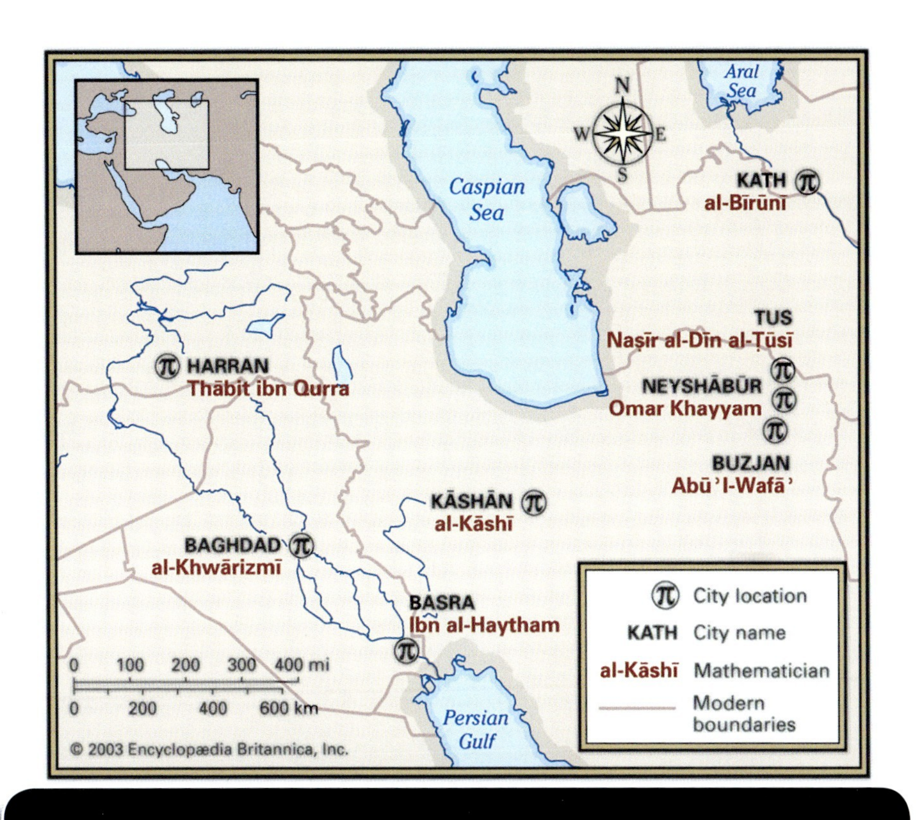

This map spans more than 600 years of prominent Islamic mathematicians.

for drawing conic sections and, above all, in the beautiful brass astrolabes with which they reduced to the turn of a dial the toil of calculating astronomical quantities.

Thābit ibn Qurra (836–901) had precisely the attributes required to bring the geometry of the Arabs up to the mark set by the Greeks. As a member of a religious sect close but hostile to both Jews and Christians, he knew Syriac and Greek as well as Arabic. As a money changer, he knew how to calculate. As both, he recommended himself to the Banū Mūsā, a set of mathematician brothers descended from a robber who had diversified into astrology. The Banū Mūsā directed a House of Wisdom in Baghdad sponsored by the caliph. There they presided over translations of the Greek classics. Thābit became an ornament of the House of Wisdom. He translated Archimedes and Apollonius, some of whose books now are known only in his versions. In a notable addition to Euclid, he tried valiantly to prove the parallel postulate.

Among the pieces of Greek geometrical astronomy that the Arabs made their own was the planispheric astrolabe, which incorporated one of the methods of projecting the celestial sphere onto a two-dimensional surface invented in ancient Greece. One of the desirable mathematical features of this method (the stereographic projection) is that it converts circles into either circles or straight lines, a property proved in the first pages of Apollonius's *Conics*. As Ptolemy showed in his *Planisphaerium*, the fact that the stereographic projection maps circles into circles or straight lines makes the astrolabe a very convenient instrument for reckoning time and representing the motions of celestial bodies. The earliest known Arabic astrolabes and manuals for their construction date from the 9th century. The Islamic world improved the astrolabe as an aid for determining the time for prayers, for finding the direction to Mecca, and for astrological divination.

EUROPE REDISCOVERS THE CLASSICS

Contacts among Christians, Jews, and Arabs in Catalonia brought knowledge of the astrolabe to the West before the year 1000. During the 12th century many manuals for its use and construction were translated into Latin along with geometrical works by the Banū Mūsā, Thābit, and others. Some of the achievements of the Arab geometers were rediscovered in the West after wide and close study of Euclid's *Elements*, which was translated repeatedly from the Arabic and once from the Greek in the 12th and 13th centuries. The *Elements* (Venice, 1482) was one of the first technical books ever printed. Archimedes also came West in the 12th century, in Latin translations from Greek and Arabic sources. Apollonius arrived only by bits and pieces. Ptolemy's *Almagest* appeared in Latin manuscript in 1175. Not until the humanists of the Renaissance turned their classical learning to mathematics, however, did the Greeks come out in standard printed editions in both Latin and Greek.

These texts affected their Latin readers with the strength of revelation. Europeans discovered the notion of proof, the power of generalization, and the superhuman cleverness of the Greeks; they hurried to master techniques that would enable them to improve their calendars and horoscopes, fashion better instruments, and raise Christian mathematicians to the level of the infidels. It took more than two centuries for the Europeans to make their unexpected heritage their own. By the 15th century, however, they were prepared to go beyond their sources. The most novel developments occurred where creativity was strongest, in the art of the Italian Renaissance.

Linear Perspective

The theory of linear perspective, the brainchild of the Florentine architect-engineers Filippo Brunelleschi (1377–1446) and Leon Battista Alberti (1404–72) and their followers, was to help remake geometry during the 17th century. The scheme of Brunelleschi and Alberti, as given without proofs in Alberti's De pictura (1435; *On Painting*), exploits the pyramid of rays that, according to what they had learned from the Westernized versions of the optics of Ibn al-Haytham (c. 965–1040), proceeds from the object to the painter's eye. Imagine, as Alberti directed, that the painter studies a scene through a window, using only one eye and not moving his head. He cannot know whether he looks at an external scene or at a glass painted to present to his eye the same visual pyramid. Supposing this decorated window to be the canvas, Alberti interpreted the painting-to-be as the projection of the scene in life onto a vertical plane cutting the visual pyramid. A distinctive feature of his system was the "point at infinity" at which parallel lines in the painting appear to converge.

Alberti's procedure, as developed by Piero della Francesca (c. 1410–92) and Albrecht Dürer (1471–1528), was used by many artists who wished to render perspective persuasively. At the same time, cartographers tried various projections of the sphere to accommodate the record of geographical discoveries that began in the mid-15th century with Portuguese exploration of the west coast of Africa. Coincidentally with these explorations, mapmakers recovered Ptolemy's *Geography*, in which he had recorded by latitude (sometimes near enough) and longitude (usually far off) the principal places known to him and indicated how they could be projected onto a map.

The discoveries that enlarged the known Earth did not fit easily on Ptolemy's projections. Cartographers therefore adopted the stereographic projection that had served astronomers. Several projected the Northern Hemisphere onto the Equator just as in the standard astrolabe, but the most widely used aspect, popularized in the world maps made by Gerardus Mercator's son for later editions of his father's atlas (beginning in 1595), projected points on the Earth onto a cylinder tangent to the Earth at the Equator. After cutting the cylinder along a vertical line and flattening the resulting rectangle, the result was the now-familiar Mercator map.

The intense cultivation of methods of projection by artists, architects, and cartographers during the Renaissance eventually provoked mathematicians into considering the properties of linear perspective in general. The most profound of these generalists was a sometime architect named Girard Desargues (1591–1661).

Transformation

French Circles

Desargues was a member of intersecting circles of 17th century French mathematicians worthy of Plato's

Academy of the 4th century BCE or Baghdad's House of Wisdom of the 9th century CE. They included René Descartes (1596–1650) and Pierre de Fermat (1601–65), inventors of analytic geometry; Gilles Personne de Roberval (1602–75), a pioneer in the development of the calculus; and Blaise Pascal (1623–62), a contributor to the calculus and an exponent of the principles set forth by Desargues.

Two main directions can be distinguished in Desargues's work. Like Renaissance artists, Desargues freely admitted the point at infinity into his demonstrations and showed that every set of parallel lines in a scene (apart from those parallel to the sides of the canvas) should project as converging bundles at some point on the "line at infinity" (the horizon). With the addition of points at infinity to the Euclidean plane, Desargues could frame all his propositions about straight lines without excepting parallel ones—which, like the others, now met one another, although not before "infinity." A farther-reaching matter arising from artistic perspective was the relation between projections of the same object from different points of view and different positions of the canvas. Desargues observed that neither size nor shape is generally preserved in projections, but collinearity is, and he provided an example, possibly useful to artists, in images of triangles seen from different points of view. The statement that accompanied this example became known as Desargues's theorem.

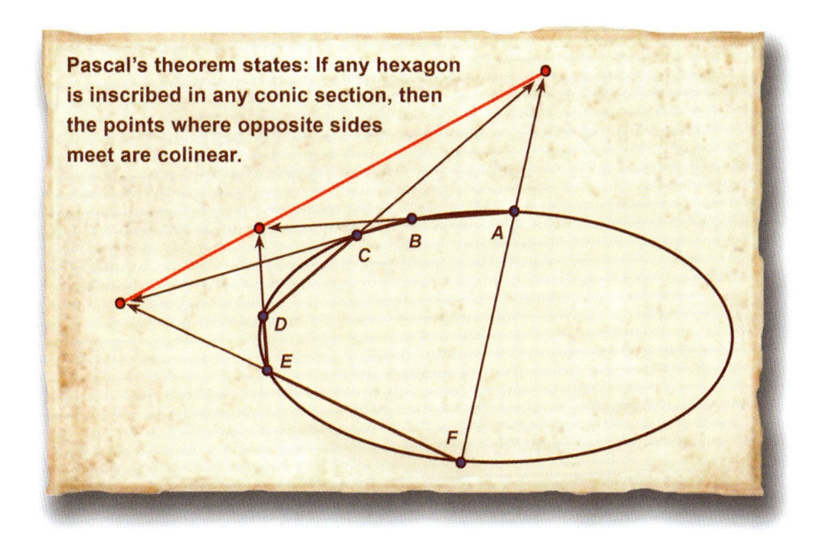

For any hexagon inscribed in any conic section, the three pairs of opposite sides when extended intersect in points that lie on a straight line.

Desargues's second direction was to "simplify" Apollonius's work on conic sections. Despite his generality of approach, Apollonius needed to prove all his theorems for each type of conic separately. Desargues saw that he could prove them all at once and, moreover, by treating a cylinder as a cone with vertex at infinity, demonstrate useful analogies between cylinders and cones. Following his lead, Pascal made his surprising discovery that the intersections of the three pairs of opposite sides of a hexagon inscribed in a conic lie on a straight line. In 1685, in his *Sectiones Conicæ*, Philippe de la Hire (1640–

1718), a Parisian painter turned mathematician, proved several hundred propositions in Apollonius's *Conics* by Desargues's efficient methods.

In 1619, as part of the great illumination that inspired Descartes to assume the modest chore of reforming philosophy as well as mathematics, he devised "compasses" made of sticks sliding in grooved frames to duplicate the cube and trisect angles. Descartes esteemed these implements and the constructions they effected as (to quote from a letter of 1619) "no less certain and geometrical than the ordinary ones with which circles are drawn." By the use of apt instruments, he would bring ancient mathematics to perfection: "scarcely anything will remain to be discovered in geometry."

What Descartes had in mind was the use of compasses with sliding members to generate curves. To classify and study such curves, Descartes took his lead from the relations Apollonius had used to classify conic sections, which contain the squares, but no higher powers, of the variables. To describe the more complicated curves produced by his instruments or defined as the loci of points satisfying involved criteria, Descartes had to include cubes and higher powers of the variables. He thus overcame what he called the deceptive character of the terms square, rectangle, and cube as used by the ancients and came to identify geometric curves as depictions of relationships defined algebraically. By reducing relations difficult to state and prove geometrically to algebraic relations between coordinates (usually rectangular) of points on curves, Descartes brought about the union of algebra and geometry that gave birth to the calculus.

GEOMETRICAL CALCULUS

The familiar use of infinity, which underlay much of perspective theory and projective geometry, also leavened the tedious Archimedean method of exhaustion. Not surprisingly, a practical man, the Flemish engineer Simon Stevin (1548–1620), who wrote on perspective and cartography among many other topics of applied mathematics, gave the first effective impulse toward redefining the object of Archimedean analysis. Instead of confining the circle between an inscribed and a circumscribed polygon, the new view regarded the circle as identical to

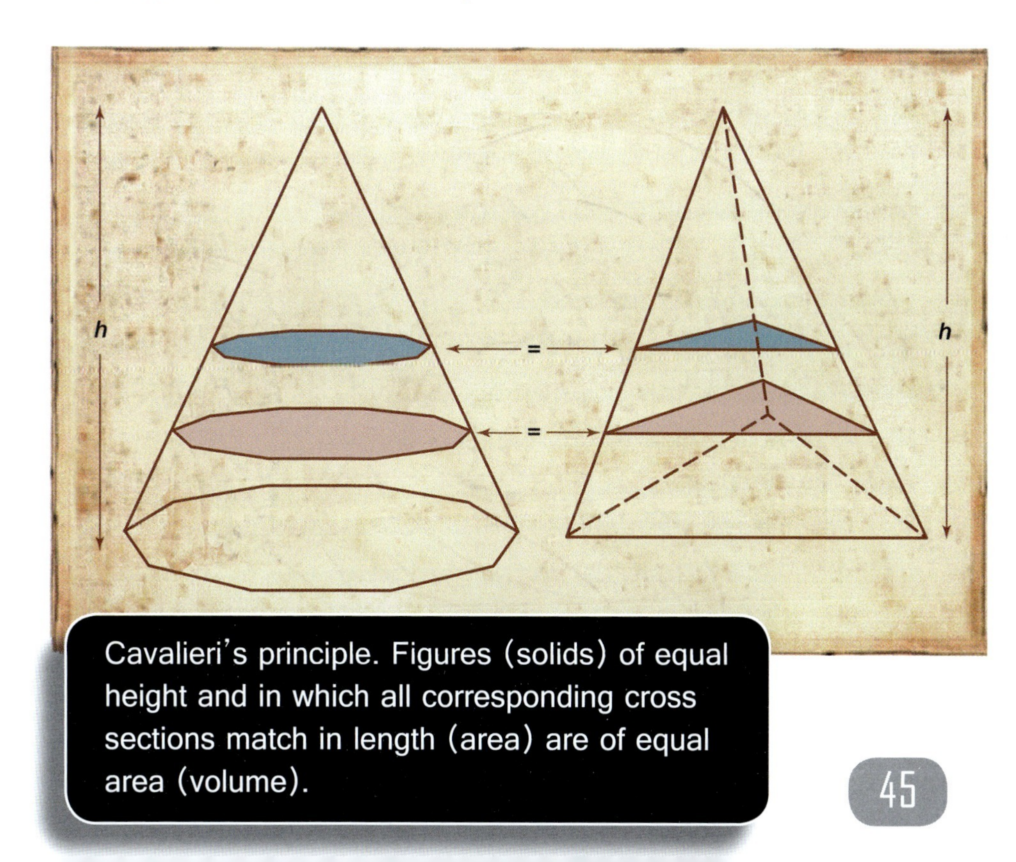

Cavalieri's principle. Figures (solids) of equal height and in which all corresponding cross sections match in length (area) are of equal area (volume).

the polygons, and the polygons to one another, when the number of their sides becomes infinitely great.

This revitalized approach to exhaustion received a preliminary systematization in the *Geometria Indivisibilibus Continuorum Nova Quadam Ratione Promota* (1635; "A Method for the Determination of a New Geometry of Continuous Indivisibles") by the Italian mathematician Bonaventura (Francesco) Cavalieri (1598–1647). Cavalieri, perhaps influenced by Kepler's method of determining volumes in *Nova Steriometria Doliorum* (1615; "New Stereometry of Wine Barrels"), regarded lines as made up of an infinite number of dimensionless points, areas as made up of lines of infinitesimal thickness, and volumes as made up of planes of infinitesimal depth in order to obtain algebraic ways of summing the elements into which he divided his figures. Cavalieri's method may be stated as follows: if two figures (solids) of equal height are cut by parallel lines (planes) such that each pair of lengths (areas) matches, then the two figures (solids) have the same area (volume). (Although not up to the rigorous standards of today and criticized by "classicist" contemporaries (who were unaware that Archimedes himself had explored similar techniques), Cavalieri's method of indivisibles became a standard tool for solving volumes until the introduction of integral calculus near the end of the 17th century.

A second geometrical inspiration for the calculus derived from efforts to define tangents to curves more complicated than conics. Fermat's method, representa-

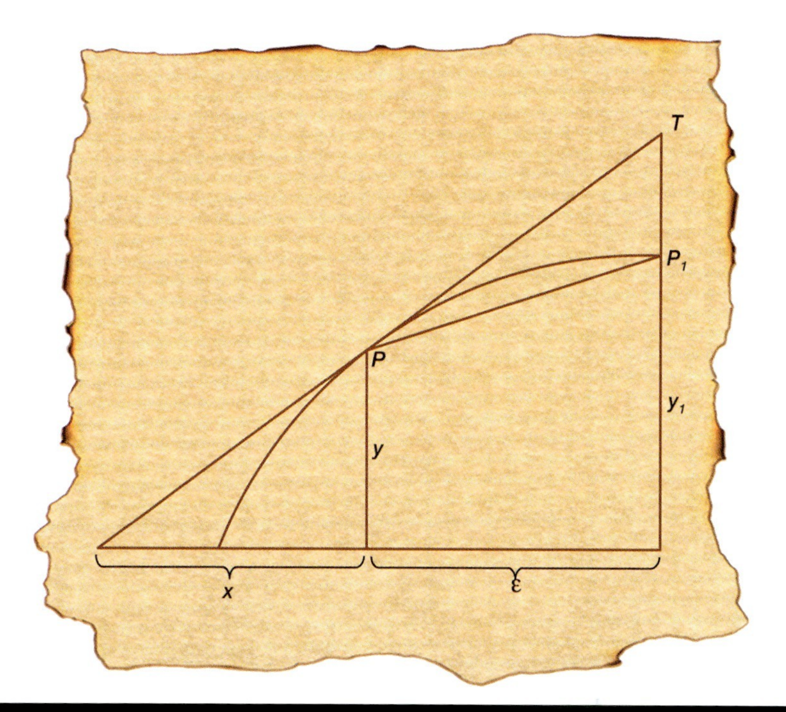

Fermat anticipated the calculus with his approach to finding the tangent line to a given curve.

tive of many, had as its exemplar the problem of finding the rectangle that maximizes the area for a given perimeter. Let the sides sought for the rectangle be denoted by a and b. Increase one side and diminish the other by a small amount ε; the resultant area is then given by $(a + \varepsilon(b - \varepsilon)$. Fermat observed what Kepler had perceived earlier in investigating the most useful shapes for wine casks, that near its maximum (or minimum) a quantity scarcely changes as the variables on which it depends alter slightly.

On this principle, Fermat equated the areas ab and $(a + \varepsilon)(b - \varepsilon)$ to obtain the stationary values: $ab = ab - \varepsilon a + \varepsilon b - \varepsilon^2$. By canceling the common term ab, dividing by ε, and then setting ε at zero, Fermat had his well-known answer, $a = b$. The figure with maximum area is a square. To obtain the tangent to a curve by this method, Fermat began with a secant through two points a short distance apart and let the distance vanish.

THE WORLD SYSTEM

Part of the motivation for the close study of Apollonius during the 17th century was the application of conic sections to astronomy. Kepler not only replaced the many circles of the old planetary system with a few ellipses, he also substituted a complicated rule of motion (his "second law") for the relatively simple Ptolemaic rule that all motions must be compounded of rotations performed at constant velocity. Kepler's second law states that a planet moves in its ellipse so that the line between it and the Sun placed at a focus sweeps out equal areas in equal times. His astronomy thus made pressing and practical the otherwise merely difficult problem of the quadrature of conics and the associated theory of indivisibles.

With the methods of Apollonius and a few infinitesimals, an inspired geometer showed that the laws regarding both area and ellipse can be derived from the suppositions that bodies free from all forces either rest or travel uniformly in straight lines and that each planet constantly falls toward the Sun with an acceleration that depends only on the distance between their centres. The inspired geometer was Isaac Newton (1642 [Old Style]–1727), who made planetary dynamics a matter entirely of geometry by replacing the planetary orbit by a succession of infinitesimal chords, planetary acceleration by a series of centripetal jerks, and, in keeping with Kepler's second law, time by an area.

Besides the problem of planetary motion, questions in optics pushed 17th-century natural philosophers and mathematicians to the study of conic sections. As Archimedes is supposed to have shown (or shone) in his destruction of a Roman fleet by reflected sunlight, a parabolic mirror brings all rays parallel to its axis to a common focus. The story of Archimedes provoked many later geometers, including Newton, to emulation. Eventually they created instruments powerful enough to melt iron.

The figuring of telescope lenses likewise strengthened interest in conics after Galileo Galilei's revolutionary improvements to the astronomical telescope in 1609. Descartes emphasized the desirability of lenses with hyperbolic surfaces, which focus bundles of parallel rays to a point (spherical lenses of wide apertures give a blurry image), and he invented a machine to cut them—which, however, proved more ingenious than useful.

A final example of early modern applications of geometry to the physical world is the old problem of the size of the Earth. On the hypothesis that the Earth cooled from a spinning liquid blob, Newton calculated that it is an oblate spheroid (obtained by rotating an ellipse around its minor axis), not a sphere, and he gave the excess of its equatorial over its polar diameter. During the 18th century many geodesists tried to find the eccentricity of the terrestrial ellipse. At first it appeared that all the measurements might be compatible with a Newtonian Earth. By the end of the century, however, geodesists had uncovered by geometry that the Earth does not, in fact, have a regular geometrical shape.

RELAXATION AND RIGOUR

The dominance of analysis (algebra and the calculus) during the 18th century produced a reaction in favour of geometry early in the 19th century. Fundamental new branches of the subject resulted that deepened, generalized, and violated principles of ancient geometry. The cultivators of these new fields, such as Jean-Victor Poncelet (1788–1867) and his self-taught disciple Jakob Steiner (1796–1863), vehemently urged the claims of geometry over analysis. The early 19th-century revival of pure geometry produced the discovery that Euclid had devoted his efforts to only one of several comprehensive geometries, the others of which can be created by replacing Euclid's fifth postulate with another about parallels.

Projection Again

Poncelet, who was an officer in the French corps of engineers, learned scraps of Desargues's work from his teacher Gaspard Monge (1746–1818), who developed his own method of projection for drawings of buildings and machines. Poncelet relied on this information to keep himself alive. Taken captive during Napoleon's invasion of Russia in 1812, he passed his time by rehearsing in his head the things he had learned from Monge. The result was projective geometry.

Poncelet employed three basic tools. One he took from Desargues: the demonstration of difficult theorems about a complicated figure by working out equivalent simpler theorems on an elementary figure interchangeable with the original figure by projection. The second tool, continuity, allows the geometer to claim certain things as true for one figure that are true of another equally general figure provided that the figures can be derived from one another by a certain process of continual change. Poncelet and his defender Michel Chasles (1793–1880) extended the principle of continuity into the domain of the imagination by considering constructs such as the common chord in two circles that do not intersect.

Poncelet's third tool was the "principle of duality," which interchanges various concepts such as points with lines, or lines with planes, so as to generate new theorems from old theorems. Desargues's theorem allows their interchange. So, as Steiner showed, does Pascal's theorem that the three points of intersection of the opposite sides of a hexagon inscribed in a conic lie on a line. Thus, the lines joining the opposite vertices of a hexagon circumscribed about a conic meet in a point.

Poncelet's followers realized that they were hampering themselves, and disguising the true fundamentality of projective geometry, by retaining the concept of length and congruence in their formulations, since projections do not usually preserve them. Similarly, parallelism had to go. Efforts were well under way by the middle of the 19th century, by Karl George Christian von Staudt (1798–1867) among others, to purge projective geometry of the last superfluous relics from its Euclidean past.

Non-Euclidean Geometries

The Enlightenment was not so preoccupied with analysis as to completely ignore the problem of Euclid's fifth postulate. In 1733 Girolamo Saccheri (1667–1733), a Jesuit professor of mathematics at the University of Pavia, Italy, substantially advanced the age-old discussion by setting forth the alternatives in great clarity and detail before declaring that he had "cleared Euclid of every defect" (*Euclides ab Omni Naevo Vindicatus*, 1733). Euclid's fifth postulate runs: "If a straight line falling on two straight lines makes the interior angles on the same side less than two right angles, the straight lines, if produced indefinitely, will meet on that side on which are the angles less than two right angles." Saccheri

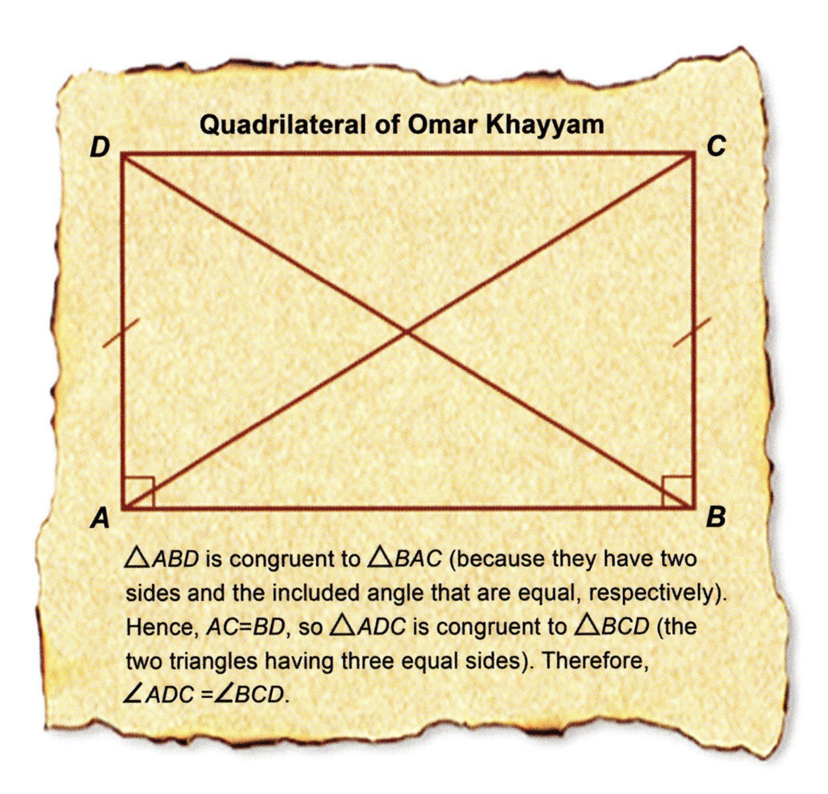

Omar Khayyam constructed the quadrilateral in an effort to prove that Euclid's fifth postulate, concerning parallel lines, is superfluous.

took up the quadrilateral of Omar Khayyam (1048–1131), who started with two parallel lines *AB* and *DC*, formed the sides by drawing lines *AD* and *BC* perpendicular to *AB*, and then considered three hypotheses for the internal angles at *C* and *D*: to be right, obtuse, or acute. The first possibility gives Euclidean geometry. Saccheri devoted himself to proving that the obtuse and the acute alternatives both end in contradictions, which would thereby eliminate the need for an explicit parallel postulate.

On the way to this spurious demonstration, Saccheri established several theorems of non-Euclidean geometry—for example, that according to whether the right, obtuse, or acute hypothesis is true, the sum of the angles of a triangle respectively equals, exceeds, or falls short of 180°. He then destroyed the obtuse hypothesis by an argument that depended upon allowing lines to increase in length indefinitely. If this is disallowed, the hypothesis of the obtuse angle produces a system equivalent to standard spherical geometry, the geometry of figures drawn on the surface of a sphere.

As for the acute angle, Saccheri could defeat it only by appealing to an arbitrary hypothesis about the behaviour of lines at infinity. One of his followers, the Swiss-German polymath Johann Heinrich Lambert (1728–77), observed that, based on the acute hypothesis, the area of a triangle is the negative of that of a spherical triangle. Since the latter is proportional to the square of the radius, r, the former appeared to Lambert to be the area of an imaginary sphere with radius ir, where $i = \sqrt{-1}$.

Although both Saccheri and Lambert aimed to establish the hypothesis of the right angle, their arguments seemed rather to indicate the unimpeachability of the alternatives. Several mathematicians at the University of Göttingen, notably the great Carl Friedrich

Gauss (1777–1855), then took up the problem. Gauss was probably the first to perceive that a consistent geometry could be built up independent of Euclid's fifth postulate, and he derived many relevant propositions, which, however, he promulgated only in his teaching and correspondence. The earliest published non-Euclidean geometric systems were the independent work of two young men from the East who had nothing to lose by their boldness. Both can be considered Gauss's disciples once removed: the Russian Nikolay Ivanovich Lobachevsky (1792–1856), who learned his mathematics from a close friend of Gauss's at the University of Kazan, where Lobachevsky later became a professor; and János Bolyai (1802–60), an officer in the Austro-Hungarian army whose father also was a friend of Gauss's. Both Lobachevsky and Bolyai had worked out their novel geometries by 1826.

Lobachevsky and Bolyai reasoned about the hypothesis of the acute angle in the manner of Saccheri and Lambert and recovered their results about the areas of triangles. They advanced beyond Saccheri and Lambert by deriving an imaginary trigonometry to go with their imaginary geometry. Just as Desargues's projective geometry was neglected for many years, so the work of Bolyai and Lobachevsky made little impression on mathematicians for a generation and more. It was largely the posthumous publication in 1855 of Gauss's ideas about non-Euclidean geometry that gave the new approaches the cachet to attract the attention of later mathematicians.

A GRAND SYNTHESIS

Another of the profound impulses Gauss gave geometry concerned the general description of surfaces. Typically—with the notable exception of the geometry of the sphere—mathematicians had treated surfaces as structures in three-dimensional Euclidean space. However, as these surfaces occupy only two dimensions, only two variables are needed to describe them. This prompted the thought that two-dimensional surfaces could be considered as "spaces" with their own geometries, not just as Euclidean structures in ordinary space. For example, the shortest distance, or path, between two points on the surface of a sphere is the lesser arc of the great circle joining them, whereas, considered as points in three-dimensional space, the shortest distance between them is an ordinary straight line.

The shortest path between two points on a surface lying wholly within that surface is called a geodesic, which reflects the origin of the concept in geodesy, in which Gauss took an active interest. His initiative in the study of surfaces as spaces and geodesics as their "lines" was pursued by his student and, briefly, his successor at Göttingen, Bernhard Riemann (1826–66). Riemann began with an abstract space of *n* dimensions. That was in the 1850s, when mathematicians and mathematical physicists were beginning to use *n*-dimensional Euclidean space to describe the motions of systems of particles in the then-new kinetic theory of gases. Riemann worked in a quasi-Euclidean space—"quasi" because he used the calculus to generalize the Pythagorean theorem to supply sufficient flexibility to provide for geodesics on any surface.

When this very general differential geometry came down to two-dimensional surfaces of constant curvature, it revealed excellent models for non-Euclidean geometries. Riemann himself pointed out that merely by calling the geodesics of a sphere "straight lines," the maligned hypoth-

The pseudosphere has constant negative curvature; i.e., it maintains a constant concavity over its entire surface.

esis of the obtuse angle produces the geometry appropriate to the sphere's surface. Similarly, as shown by Eugenio Beltrami (1835–1900), who ended his teaching career in Saccheri's old post at Pavia, the geometry defined in the plane by the hypothesis of the acute angle fits perfectly a surface of revolution of constant negative curvature now called a pseudosphere—again, provided that its geodesics are accepted as the straight lines of the geometry.

Since the hypothesis of the obtuse angle correctly characterizes Euclidean geometry applied to the surface of a sphere, the non-Euclidean geometry based on it must be exactly as consistent as Euclidean geometry. The case of the acute angle treated by Lobachevsky and Bolyai required a sharper tool. Beltrami found it in a projection into a disc in the Euclidean plane of the points of a non-Euclidean space, in which each geodesic from the non-Euclidean space corresponds to a chord of the disc. Geometry built on the hypothesis of the acute angle has the same consistency as Euclidean geometry.

The key role of Euclidean geometry in proofs of the consistency of non-Euclidean geometries exposed the *Elements* to ever-deeper scrutiny. The old blemishes—particularly appeals to intuition and diagrams for the meaning of concepts like "inside" and "between" and the use of questionable procedures like superposition to prove congruency—became intolerable to mathematicians who laboured to clarify the foundations of arithmetic and the calculus as well as the interrelations of the new geometries.

The German mathematician Moritz Pasch (1843–1930), in his *Vorlesungen über neuere Geometrie* (1882; "Lectures on the New Geometry"), identified what was wanting: undefined concepts, axioms about those concepts, and more rigorous logic based on those axioms. The choice of undefined concepts and axioms is free, apart from the constraint of consistency. Mathematicians following Pasch's path introduced various elements and axioms and developed their geometries with greater or lesser elegance and trouble. The most successful of these systematizers was the Göttingen professor David Hilbert (1862–1943), whose *The Foundations of Geometry* (1899) greatly influenced efforts to axiomatize all of mathematics.

The Real World

Euclid's *Elements* had claimed the excellence of being a true account of space. Within this interpretation, Euclid's fifth postulate was an empirical finding, while non-Euclidean geometries did not apply to the real world. Bolyai apparently could not free himself from the persuasion that Euclidean geometry represented reality. Lobachevsky observed that, if there were a star so distant that its parallax was not observable from the Earth's orbit, his geometry would be indistinguishable from Euclid's at the point where the parallax vanished. By his calculation, based on stellar parallaxes then just detected, his geometry could be physically meaningful only in gargantuan triangles spanning interstellar space.

In fact, non-Euclidean geometries apply to the cosmos more locally than Lobachevsky imagined. In 1916 Albert Einstein (1879–1955) published "The Foundation of the General Theory of Relativity," which replaced Newton's description of gravitation as a force that attracts distant masses to each other through Euclidean space with a principle of least effort, or shortest (temporal) path, for motion along the geodesics of a curved space. Einstein not only explained how gravitating bodies give this surface its properties—that is, mass determines how the differential distances, or curvatures, in Riemann's geometry differ from those in Euclidean space—but also successfully predicted the deflection of light, which has no mass, in the vicinity of a star or other massive body. This was an extravagant piece of geometrizing—the replacement of gravitational force by the curvature of a surface. But it was not all. In relativity theory time is considered to be a dimension along with the three dimensions of space. On the closed four-dimensional world thus formed, the history of the universe stands revealed as describable by motion within a vast congeries of geodesics in a non-Euclidean universe.

Chapter 2

BRANCHES OF GEOMETRY

Several ancient cultures had developed a form of geometry suited to the relationships among lengths, areas, and volumes of physical objects. This geometry was codified in Euclid's *Elements* about 300 BCE on the basis of 10 axioms, or postulates, from which several hundred theorems were proved by deductive logic. The *Elements* epitomized the axiomatic-deductive method for many centuries.

Euclidean Geometry

In its rough outline, Euclidean geometry is the plane and solid geometry commonly taught in secondary schools. Indeed, until the second half of the 19th century, when non-Euclidean geometries attracted the attention of mathematicians, geometry meant Euclidean geometry. It is the most typical expression of general mathematical thinking. Rather than the memorization of simple algorithms to solve equations by rote, it demands true insight into the subject, clever ideas for applying theorems in special situations, an ability to generalize from known facts, and an insistence on the importance of proof. In Euclid's great work, the *Elements*, the only tools employed for geometrical constructions were the ruler and compass—a restriction retained in elementary Euclidean geometry to this day.

In its rigorous deductive organization, the *Elements* remained the very model of scientific exposition until the end of the 19th century, when the German mathematician David Hilbert wrote his famous *Foundations of Geometry* (1899). The modern version of Euclidean geometry is the theory of Euclidean (coordinate) spaces of multiple dimensions, where distance is measured by a suitable generalization of the Pythagorean theorem.

FUNDAMENTALS

Euclid realized that a rigorous development of geometry must start with the foundations. Hence, he began the *Elements* with some undefined terms, such as "a point is that which has no part" and "a line is a length without breadth." Proceeding from these terms, he defined further ideas such as angles, circles, triangles, and various other polygons and figures. For example, an angle was defined as the inclination of two straight lines, and a circle was a plane figure consisting of all points that have a fixed distance (radius) from a given centre.

As a basis for further logical deductions, Euclid proposed five common notions, such as "things equal to the same thing are equal," and five unprovable but intuitive principles known variously as postulates or axioms. Stated in modern terms, the axioms are as follows:

1. Given two points, there is a straight line that joins them.
2. A straight line segment can be prolonged indefinitely.
3. A circle can be constructed when a point for its centre and a distance for its radius are given.
4. All right angles are equal.
5. If a straight line falling on two straight lines makes the interior angles on the same side less than two right angles, the two straight lines, if produced indefinitely, will meet on that side on which the angles are less than the two right angles.

Hilbert refined axioms (1) and (5) as follows:

1. For any two different points, (a) there exists a line containing these two points, and (b) this line is unique.
5. For any line L and point p not on L, (a) there exists a line through p not meeting L, and (b) this line is unique.

The fifth axiom became known as the "parallel postulate," since it provided a basis for the uniqueness of parallel lines. (It also attracted great interest because it seemed less intuitive or self-evident than the others. In the 19th century, Carl Friedrich Gauss, János Bolyai, and Nikolay Lobachevsky all began to experiment with this postulate, eventually arriving at new, non-Euclidean, geometries.) All five axioms provided the basis for numerous provable statements, or theorems, on which Euclid built his geometry. The following sections briefly explain the most important theorems of Euclidean plane and solid geometry.

PLANE GEOMETRY

Two triangles are said to be congruent if one can be exactly superimposed on the other by a rigid motion, and the congruence theorems specify the conditions under which this can occur. One of the first congruences is the side-angle-side (SAS) theorem: If two sides and the included angle of one triangle are equal to two sides and the included angle of another triangle, the triangles are congruent. Following this, there are corresponding angle-side-angle (ASA) and side-side-side (SSS) theorems.

The first very useful theorem derived from the axioms is the basic symmetry property of isosceles triangles—that is, that two sides of a triangle are equal if and only if the angles opposite them are equal. Euclid's proof of this theorem was once called Pons Asinorum ("Bridge of Asses"), supposedly because mediocre students could not proceed across it to the farther reaches

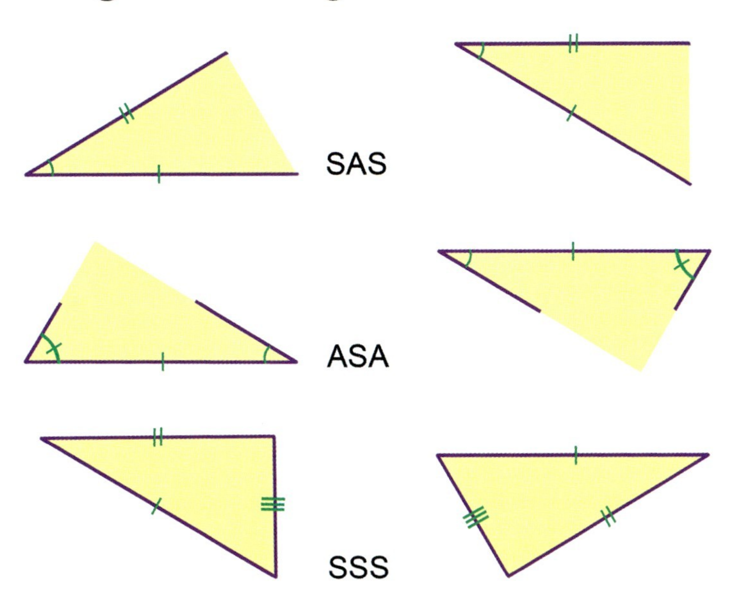

Triangles are congruent if: two sides and the included angle are equal (SAS); two angles and the included side are equal (ASA); or all three sides are equal (SSS).

of geometry. The Bridge of Asses opens the way to various theorems on the congruence of triangles.

The parallel postulate is fundamental for the proof of the theorem, attributed to the Pythagoreans, that the sum of the angles of a triangle is always 180 degrees.

As indicated above, congruent figures have the same shape and size. Similar figures, on the other hand, have the same shape but may differ in size.

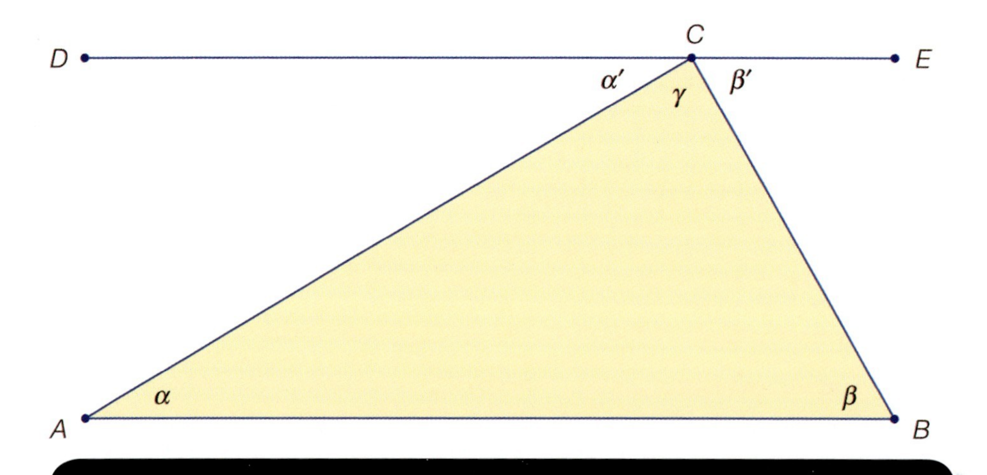

Proof that the sum of the angles in a triangle is 180 degrees.

Shape is intimately related to the notion of proportion, as ancient Egyptian artisans observed long ago. Segments of lengths a, b, c, and d are said to be proportional if $a:b = c:d$ (read, a is to b as c is to d; in older notation $a:b::c:d$). The fundamental theorem of similarity states that a line segment splits two sides of a triangle into proportional segments if and only if the segment is parallel to the triangle's third side.

The similarity theorem may be reformulated as the AAA (angle-angle-angle) similarity theorem: two triangles have their corresponding angles equal if and only if their corresponding sides are proportional. Two similar triangles are related by a scaling (or similarity) factor s: if the first triangle has sides a, b, and c, then the second one will have sides sa, sb, and sc. In addition to the ubiquitous use of scaling factors on construction plans and geographic maps, similarity is fundamental to trigonometry.

Just as a segment can be measured by comparing it with a unit segment, the area of a polygon or other plane figure can be measured by comparing it with a unit square. The common formulas for calculating areas reduce this kind of measurement to the measurement of certain suitable lengths. The simplest case is a rectangle with sides *a* and *b*, which has area *ab*. By putting a triangle into an appropriate rectangle, one can show that the area of the triangle is half the product of the length of one of its bases and its corresponding height—

Fundamental theorem of similarity

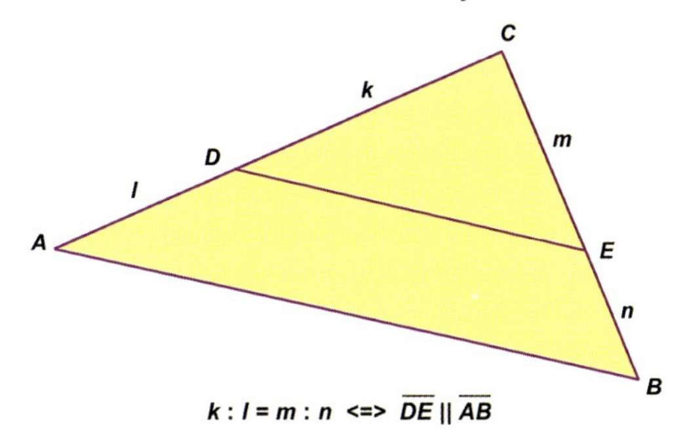

This theorem then enables one to show that the small and large triangles are similar.

Proof that the area of a triangle = ½ base • height

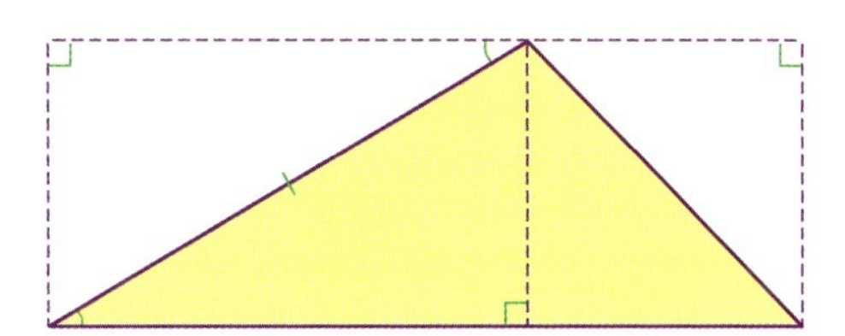

The right triangle △AFB is ½ of the rectangle □*ADBF*.

Similarly, △*BFC* is ½ of □*BECF*.

Thus, the area of △*ABC* = ½ area of □*ADEC* = ½ *AC* • *BF* = ½ *bh*.

Area of a triangle.

$bh/2$. One can then compute the area of a general polygon by dissecting it into triangular regions. If a triangle (or more general figure) has area A, a similar triangle (or figure) with a scaling factor of s will have an area of s^2A.

PYTHAGOREAN THEOREM

For a triangle $\triangle ABC$ the Pythagorean theorem has two parts: (1) if $\angle ACB$ is a right angle, then $a^2 + b^2 = c^2$; (2) if $a^2 + b^2 = c^2$, then $\angle ACB$ is a right angle. For an arbitrary triangle, the Pythagorean theorem is generalized to the law of cosines: $a^2 + b^2 = c^2 - 2ab \cos(\angle ACB)$. When $\angle ACB$ is 90 degrees, this reduces to the Pythagorean theorem because $\cos(90°) = 0$.

Since Euclid, a host of professional and amateur mathematicians have found more than 300 distinct proofs of the Pythagorean theorem. Despite its antiquity, it remains one of the most important theorems in mathematics. It enables one to calculate distances or, more important, to define distances in situations far more general than elementary geometry. For example, it has been generalized to multidimensional vector spaces.

CIRCLES

A chord AB is a segment in the interior of a circle connecting two points (A and B) on the circumference. When a chord passes through the circle's centre, it is a diameter, d. The circumference of a circle is given by πd, or $2\pi r$ where r is the radius of the circle; the area of a circle is πr^2. In each case, π is the same constant (3.14159…). The Greek mathematician Archimedes (c. 287–212/211 BCE) used the method of exhaustion to obtain upper and lower bounds for π by circumscribing and inscribing regular polygons about a circle.

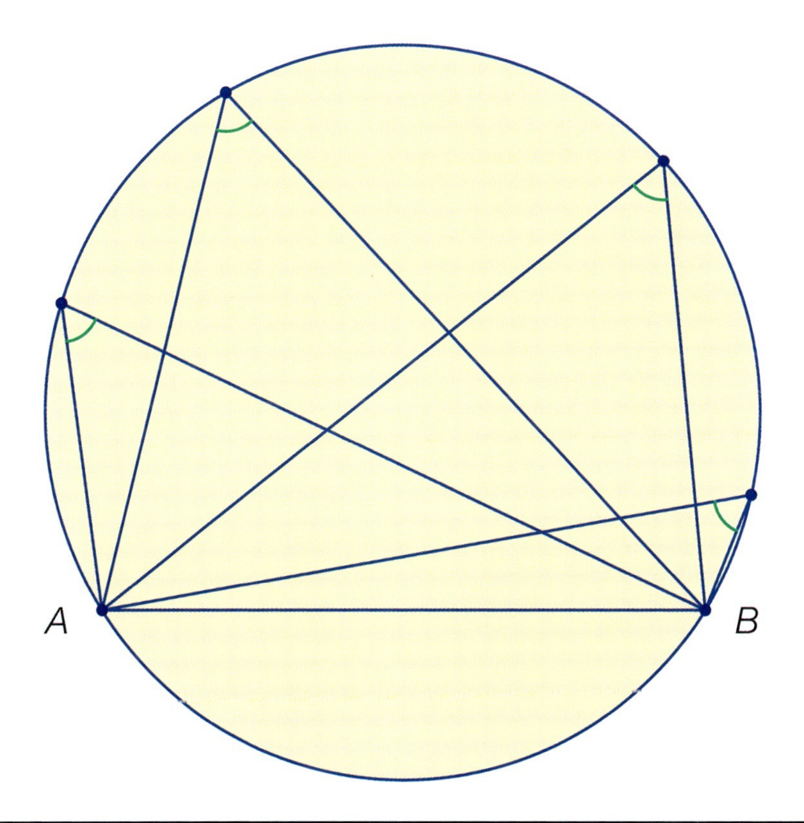

Thales of Miletus is generally credited with giving the first proof that for any chord *AB* in a circle, all of the angles subtended by points anywhere on the same semiarc of the circle will be equal.

A semicircle has its end points on a diameter of a circle. Thales (flourished 6th century BCE) is generally credited with having proved that any angle inscribed in a semicircle is a right angle; that is, for any point C on the semicircle with diameter AB, $\angle ACB$ will always be 90 degrees. Another important theorem states that for any chord AB in a circle, the angle subtended by any point on the same semiarc of the circle will be invariant. Slightly modified, this means that in a circle, equal chords determine equal angles, and vice versa.

Summarizing the above material, the five most important theorems of plane Euclidean geometry are: the sum of the angles in a triangle is 180 degrees, the Bridge of Asses, the fundamental theorem of similarity, the Pythagorean theorem, and the invariance of angles subtended by a chord in a circle. Most of the more advanced theorems of plane Euclidean geometry are proven with the help of these theorems.

REGULAR POLYGONS

A polygon is called regular if it has equal sides and angles. Thus, a regular triangle is an equilateral triangle, and a regular quadrilateral is a square. A general problem since antiquity has been the problem of constructing a regular n-gon, for different n, with only ruler and compass. For example, Euclid constructed a regular pentagon by applying the above-mentioned five important theorems in an ingenious combination.

Techniques, such as bisecting the angles of known constructions, exist for constructing regular n-gons for many values, but none is known for the general case. In 1797, following centuries without any progress, Gauss surprised the mathematical community by discovering a construction for the 17-gon. More generally, Gauss was able to show that for a prime number p, the regular p-gon is constructible if and only if p is a "Fermat prime": $p = F(k) = 2^{2^k} + 1$. Because it is not known in general which $F(k)$ are prime, the construction problem for regular n-gons is still open.

Three other unsolved construction problems from antiquity were finally settled in the 19th century by applying tools not available to the Greeks. Comparatively simple algebraic methods showed that it is not possible to trisect an angle with ruler and compass or to construct a cube with a volume double that of a given cube. To show that it is not possible to square a circle (i.e., to construct a square equal in area to a given circle by the same means), however, demanded deeper insights into the nature of the number π.

CONIC SECTIONS AND GEOMETRIC ART

The most advanced part of plane Euclidean geometry is the theory of the conic sections (the ellipse, parabola, and hyperbola). Much as the *Elements* displaced all other introductions to geometry, the *Conics* of Apollonius of Perga (c. 240–190 BCE), known by his contemporaries as "The Great Geometer," was for many centuries the definitive treatise on the subject.

Medieval Islamic artists explored ways of using geometric figures for decoration. For example, the decorations of the Alhambra of Granada, Spain, demonstrate an understanding of all 17 of the different "Wallpaper groups" that can be used to tile the plane. In the 20th century, internationally renowned artists such as Josef Albers, Max Bill, and Sol Le Witt were inspired by motifs from Euclidean geometry.

SOLID GEOMETRY

The most important difference between plane and solid Euclidean geometry is that human beings can look at the plane "from above," whereas three-dimensional space cannot be looked at "from outside." Consequently, intuitive insights are more difficult to obtain for solid geometry than for plane geometry.

Some concepts, such as proportions and angles, remain unchanged from plane to solid geometry. For other familiar concepts, there exist analogies—most noticeably, volume for area and three-dimensional shapes for two-dimensional shapes (sphere for circle, tetrahedron for triangle, box for rectangle). However, the theory of tetrahedra is not nearly as rich as it is for triangles. Active research in higher-dimensional Euclidean geometry includes convexity and sphere packings and their applications in cryptology and crystallography.

VOLUME

As explained above, in plane geometry the area of any polygon can be calculated by dissecting it into triangles. A similar procedure is not possible for solids. In 1901 the German mathematician Max Dehn showed that there exist a cube and a tetrahedron of equal volume that cannot be dissected and rearranged into each other. This means that calculus must be used to calculate volumes for even many simple solids like pyramids.

REGULAR SOLIDS

Regular polyhedra are the solid analogies to regular polygons in the plane. Regular polygons are defined as having equal (congruent) sides and angles. In an analogy, a solid is called regular if its faces are congruent regular polygons and its polyhedral angles (angles at which the faces meet) are congruent. This concept has been generalized to higher-dimensional (coordinate) Euclidean spaces.

Whereas in the plane there exist (in theory) infinitely many regular polygons, in three-dimensional space there exist exactly five regular polyhedra. These are known as the Platonic solids: the tetrahedron, or pyramid, with 4 triangular faces; the cube, with 6 square faces; the octahedron, with 8 equilateral triangular faces; the dodecahedron, with 12 pentagonal faces; and the icosahedron, with 20 equilateral triangular faces.

In four-dimensional space there exist exactly six regular polytopes, five of them generalizations from three-dimensional space. In any space of more than four dimensions there exist exactly three regular polytopes, the generalizations of the tetrahedron, the cube, and the octahedron.

CONIC SECTION

Any curve produced by the intersection of a plane and a right circular cone is known as a conic section. Depending on the angle of the plane relative to the cone, the intersection is a circle, an ellipse, a hyperbola, or a parabola. Special (degenerate) cases of intersection occur when the plane passes through only the apex (producing a single point) or through the apex and another point on the cone (producing one straight line or two intersecting straight lines).

The basic descriptions, but not the names, of the conic sections can be traced to Menaechmus (fl. c. 350 BCE), a pupil of both Plato and Eudoxus of Cnidus. Apollonius of Perga (c. 262–190 BCE), the "Great Geometer,"

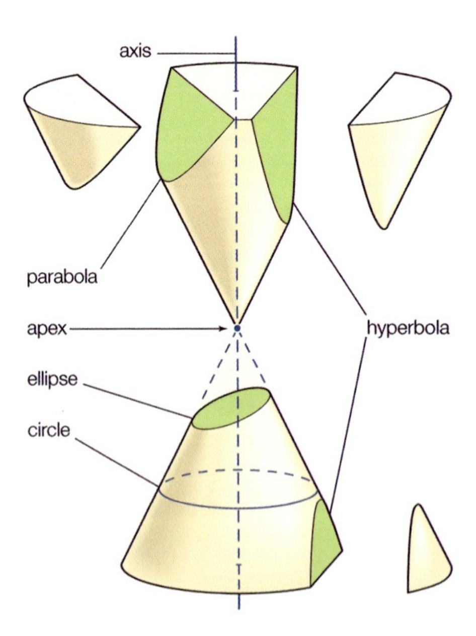

The conic sections result from intersecting a plane with a double cone.

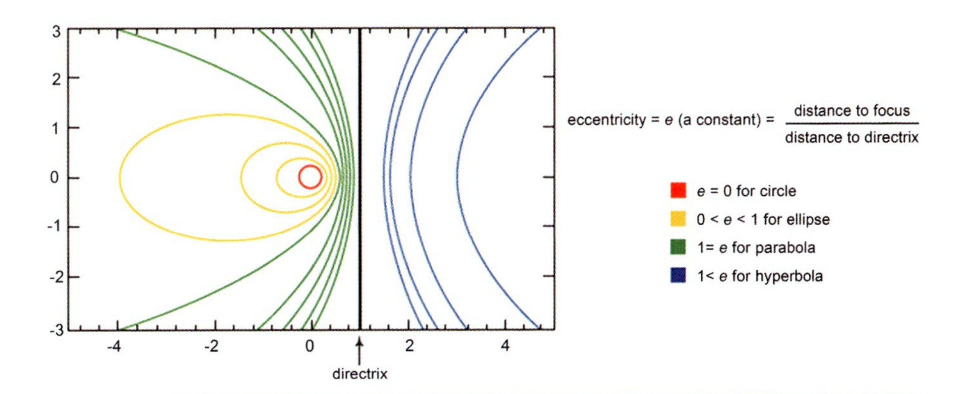

The eccentricity of a conic section completely characterizes its shape.

gave the conic sections their names and was the first to define the two branches of the hyperbola (which presuppose the double cone). Apollonius's eight-volume treatise on the conic sections, *Conics*, is one of the greatest scientific works from the ancient world.

Conics may also be described as plane curves that are the paths (loci) of a point moving so that the ratio of its distance from a fixed point (the focus) to the distance from a fixed line (the directrix) is a constant, called the eccentricity of the curve. If the eccentricity is zero, the curve is a circle; if equal to one, a parabola; if less than one, an ellipse; and if greater than one, a hyperbola.

Every conic section corresponds to the graph of a second degree polynomial equation of the form $Ax^2 + By^2 + 2Cxy + 2Dx + 2Ey + F = 0$, where x and y are variables and A, B, C, D, E, and F are coefficients that depend upon the particular conic. By a suitable choice of coordinate axes, the equation for any conic can be reduced to one of three simpler forms:

$$x^2/a^2 + y^2/b^2 = 1,\ x^2/a^2 - y^2/b^2 = 1,\ \text{or}\ y^2 = 2px,$$

corresponding to an ellipse, a hyperbola, and a parabola, respectively. (An ellipse where $a = b$ is in fact a circle.) The extensive use of coordinate systems for the algebraic analysis of geometric curves originated with René Descartes (1596–1650).

The early history of conic sections is joined to the problem of "doubling the cube." According to Eratosthenes of Cyrene (c. 276–190 BCE), the people of Delos consulted the oracle of Apollo for aid in ending a plague (c. 430 BCE) and were instructed to build Apollo a new altar of twice the old altar's volume and with the same cubic shape. Perplexed, the Delians consulted Plato, who stated that "the oracle meant, not that the god wanted an altar of double the size, but that he wished, in setting them the task, to shame the Greeks for their neglect of mathematics and their contempt for geometry." Hippocrates of Chios (c. 470–410 BCE) first discovered that the "Delian problem" can be reduced to finding two mean proportionals between a and $2a$ (the volumes of the respective altars)—that is, determining x and y such that $a{:}x = x{:}y = y{:}2a$. This is equivalent to solving simultaneously any two of the equations $x^2 = ay$, $y^2 = 2ax$, and $xy = 2a^2$, which correspond to two parabolas and a hyperbola, respectively. Later, Archimedes (c.

287–212/211 BCE) showed how to use conic sections to divide a sphere into two segments having a given ratio.

Diocles (c. 200 BCE) demonstrated geometrically that rays—for instance, from the Sun—that are parallel to the axis of a paraboloid of revolution (produced by rotating a parabola about its axis of symmetry) meet at

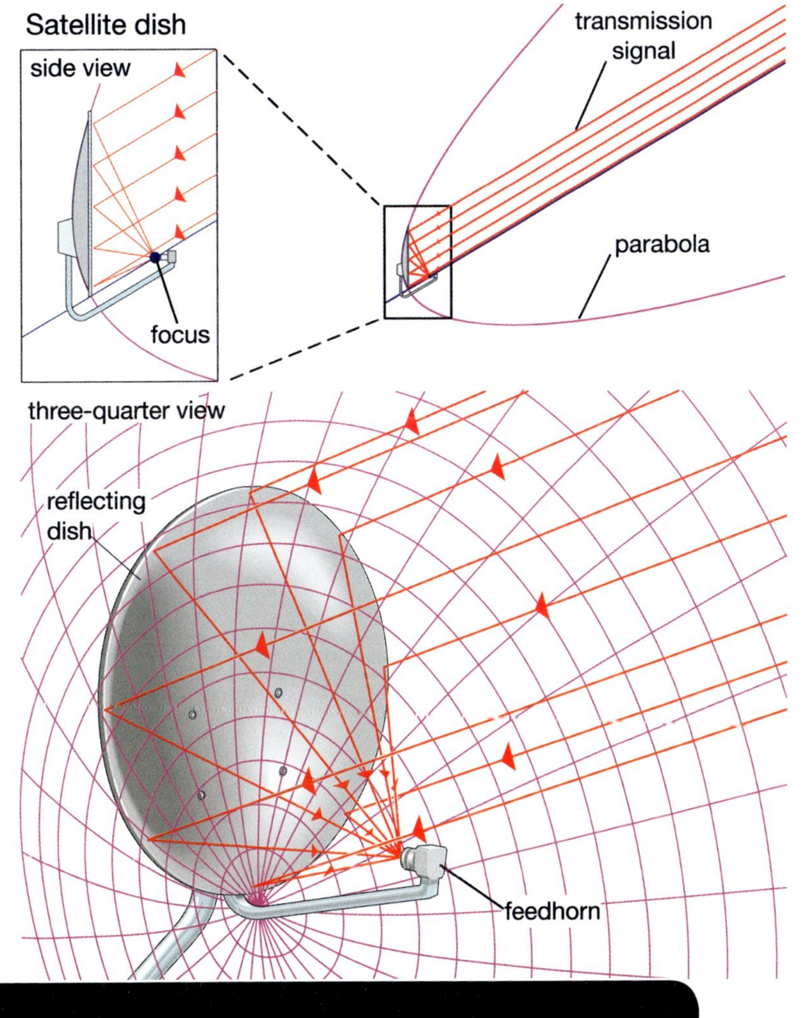

Parabolic satellite dish antenna illustrating Diocles' geometric demonstration of rays meeting at the focus.

the focus. Archimedes is said to have used this property to set enemy ships on fire. The focal properties of the ellipse were cited by Anthemius of Tralles, one of the architects for Hagia Sophia Cathedral in Constantinople (completed in 537 CE), as a means of ensuring that an altar could be illuminated by sunlight all day.

Conic sections found their first practical application outside of optics in 1609 when Johannes Kepler derived his first law of planetary motion: A planet travels in an ellipse with the Sun at one focus. Galileo Galilei published the first correct description of the path of projectiles—a parabola—in his *Dialogues of the Two New Sciences* (1638). In 1639 the French engineer Girard Desargues initiated the study of those properties of conics that are invariant under projections. Eighteenth-century architects created a fad for whispering galleries—such as in the U.S. Capital and in St. Paul's Cathedral in London—in which a whisper at one focus of an ellipsoid (an ellipse rotated about one axis) can be heard at the other focus, but nowhere else. From the ubiquitous parabolic satellite dish to the use of ultrasound in lithotripsy, new applications for conic sections continue to be found.

ANALYTIC GEOMETRY

Analytic geometry was initiated by the French mathematician René Descartes (1596–1650), who introduced rectangular coordinates to locate points and to enable lines and curves to be represented with algebraic equations. Algebraic geometry is a modern extension of the subject to multidimensional and non-Euclidean spaces.

The importance of analytic geometry is that it establishes a correspondence between geometric curves and algebraic equations. This correspondence makes it possible to reformulate problems in geometry as equivalent problems in algebra, and vice versa. The methods of either subject can then be used to solve problems in the other. For example, computers create animations for display in games and films by manipulating algebraic equations.

Elementary Analytic Geometry

Apollonius of Perga (c. 262–190 BCE) foreshadowed the development of analytic geometry by more than 1,800 years with his book *Conics*. He defined a conic as the intersection of a cone and a plane. Using Euclid's results on similar triangles and on secants of circles, he found a relation satisfied by the distances from any point P of a conic to two perpendicular lines, the major axis of the conic and the tangent at an endpoint of the axis. These distances correspond to coordinates of P, and the relation between these coordinates corresponds to a quadratic equation of the conic. Apollonius used this relation to deduce fundamental properties of conics.

Further development of coordinate systems in mathematics emerged only after algebra had matured under Islamic and Indian mathematicians. At the end of the 16th

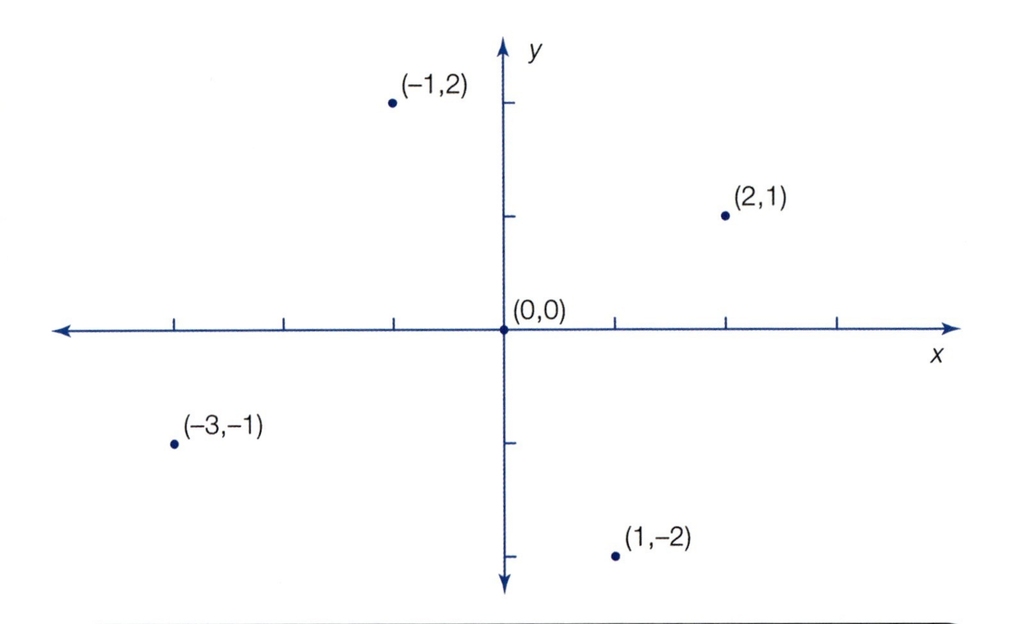

Several points are labeled in a two-dimensional graph, known as the Cartesian plane.

century, the French mathematician François Viète introduced the first systematic algebraic notation, using letters to represent known and unknown numerical quantities, and he developed powerful general methods for working with algebraic expressions and solving algebraic equations. With the power of algebraic notation, mathematicians were no longer completely dependent upon geometric figures and geometric intuition to solve problems. The more daring began to leave behind the standard geometric way of thinking in which linear (first power) variables corresponded to lengths, squares (second power) to areas, and cubics (third power) to volumes, with higher powers lacking "physical" interpretation. Two Frenchmen, the mathematician-philosopher René Descartes and the lawyer-mathematician Pierre de Fermat, were among the first to take this daring step.

Descartes and Fermat independently founded analytic geometry in the 1630s by adapting Viète's algebra to the study of geometric loci. They moved decisively beyond Viète by using letters to represent distances that are variable instead of fixed. Descartes used equations to study curves defined geometrically, and he stressed the need to consider general algebraic curves—graphs of polynomial equations in x and y of all degrees. He demonstrated his method on a classical problem: finding all points P such that the product of the distances from P to certain lines equals the product of the distances to other lines.

Fermat emphasized that any relation between x and y coordinates determines a curve. Using this idea, he recast Apollonius's arguments in algebraic terms and restored lost work. Fermat indicated that any quadratic equation in x and y can be put into the standard form of one of the conic sections.

Fermat did not publish his work, and Descartes deliberately made his hard to read in order to discourage "dabblers." Their ideas gained general acceptance only through the efforts of other mathematicians in the latter half of the 17th century. In particular, the Dutch mathematician Frans van Schooten translated Descartes's writings from French to Latin. He added vital explanatory material, as did the French lawyer Florimond de Beaune, and the Dutch mathematician Johan de Witt. In England, the mathematician John Wallis popularized analytic geometry, using equations to define conics and derive their properties. He used negative coordinates freely, although it was Isaac Newton who unequivocally used two (oblique) axes to divide the plane into four quadrants.

Analytic geometry had its greatest impact on mathematics via calculus. Without access to the power of analytic geometry, classical Greek mathematicians such as Archimedes (c. 287–212/211 BCE) solved special cases of the basic problems of calculus: finding tangents and extreme points (differential calculus) and arc lengths, areas, and volumes (integral calculus). Renaissance mathematicians were led back to these problems by the

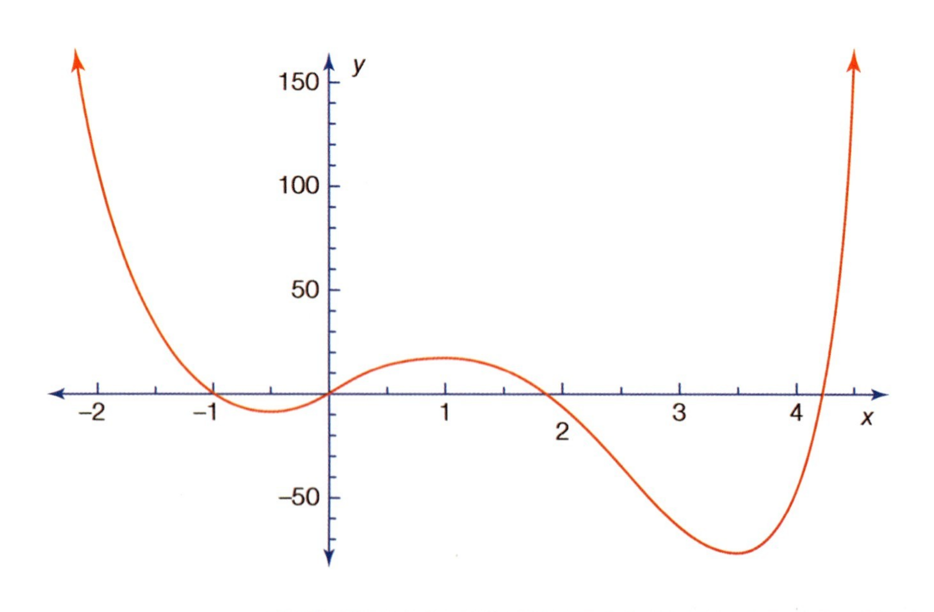

Polynomial graph.

needs of astronomy, optics, navigation, warfare, and commerce. They naturally sought to use the power of algebra to define and analyze a growing range of curves.

Fermat developed an algebraic algorithm for finding the tangent to an algebraic curve at a point by finding a line that has a double intersection with the curve at the point—in essence, inventing differential calculus. Descartes introduced a similar but more complicated algorithm using a circle. Fermat computed areas under the curves $y = ax^k$ for all rational numbers $k \neq -1$ by summing areas of inscribed and circumscribed rectangles. For the rest of the 17th century, the groundwork for calculus was continued by many mathematicians, including the Frenchman Gilles Personne de Roberval, the Italian Bonaventura Cavalieri, and the Britons James Gregory, John Wallis, and Isaac Barrow.

Newton and the German Gottfried Leibniz revolutionized mathematics at the end of the 17th century by independently demonstrating the power of calculus. Both men used coordinates to develop notations that expressed the ideas of calculus in full generality and led naturally to differentiation rules and the fundamental theorem of calculus (connecting differential and integral calculus).

Newton demonstrated the importance of analytic methods in geometry, apart from their role in calculus, when he asserted that any cubic—or, algebraic curve of degree three—has one of four standard equations,

$$xy^2 + ey = ax^3 + bx^2 + cx + d,$$

$$xy = ax^3 + bx^2 + cx + d,$$

$$y^2 = ax^3 + bx^2 + cx + d,$$

$$y = ax^3 + bx^2 + cx + d,$$

for suitable coordinate axes. The Scottish mathematician James Stirling proved this assertion in 1717, possibly with Newton's aid. Newton divided cubics into 72 species, a total later corrected to 78.

Newton also showed how to express an algebraic curve near the origin in terms of the fractional power series $y = a_1x^{1/k} + a_2x^{2/k} + \ldots$ for a positive integer k. Mathematicians have since used this technique to study algebraic curves of all degrees.

ANALYTIC GEOMETRY OF THREE AND MORE DIMENSIONS

Although both Descartes and Fermat suggested using three coordinates to study curves and surfaces in space, three-dimensional analytic geometry developed slowly until about 1730, when the Swiss mathematicians Leonhard Euler and Jakob Hermann and the French mathematician Alexis Clairaut produced general equations for cylinders, cones, and surfaces of revolution. For example, Euler and Hermann showed that the equation $f(z) = x^2 + y^2$ gives the surface that is produced by revolving the curve $f(z) = x^2$ about the z-axis.

Newton made the remarkable claim that all plane cubics arise from those in his third standard form by pro-

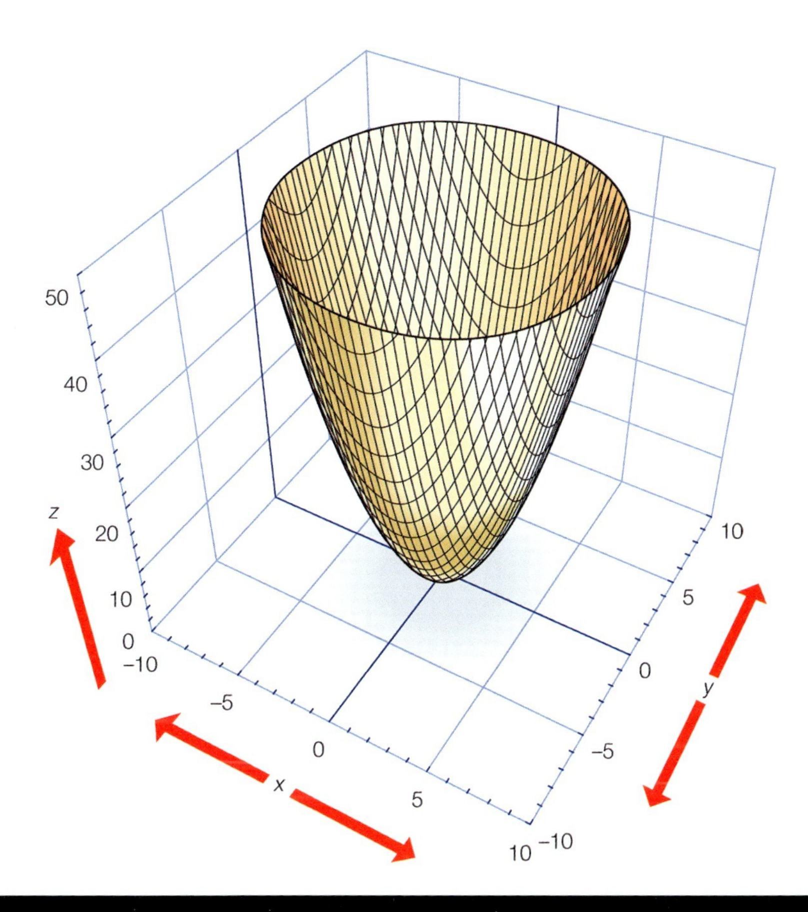

Elliptic paraboloid.

jection between planes. This was proved independently in 1731 by Clairaut and the French mathematician François Nicole. Clairaut obtained all the cubics in Newton's four standard forms as sections of the cubical cone

$$zy^2 = ax^3 + bx^2z + cxz^2 + dz^3$$

consisting of the lines in space that join the origin (0,

0, 0) to the points on the third standard cubic in the plane $z = 1$.

In 1748 Euler used equations for rotations and translations in space to transform the general quadric surface

$$ax^2 + by^2 + cz^2 + dxy + exz + fyz + gx + hy + iz + j = 0$$

so that its principal axes coincide with the coordinate axes. Euler and the French mathematicians Joseph-Louis Lagrange and Gaspard Monge made analytic geometry independent of synthetic (nonanalytic) geometry.

In Euclidean space of any dimension, vectors—directed line segments—can be specified by coordinates. An n-tuple $(a_1, \ldots, a_n)$ represents the vector in n-dimensional space that projects onto the real numbers $a_1, \ldots, a_n$ on the coordinate axes.

In 1843 the Irish mathematician-astronomer William Rowan Hamilton represented four-dimensional vectors algebraically and invented the quaternions, the first non-commutative algebra to be extensively studied. Multiplying quaternions with one coordinate zero led Hamilton to discover fundamental operations on vectors. Nevertheless, mathematical physicists found the notation used in vector analysis more flexible—in particular, it is readily extendable to infinite-dimensional spaces. The quaternions remained of interest algebraically and were incorporated in the 1960s into certain new particle physics models.

As readily available computing power grew exponentially in the last decades of the 20th century, computer animation and computer-aided design became ubiquitous. These applications are based on three-dimensional analytic geometry. Coordinates are used to determine the edges or parametric curves that form boundaries of the surfaces of virtual objects. Vector analysis is used to model lighting and determine realistic shadings of surfaces.

As early as 1850, Julius Plücker had united analytic and projective geometry by introducing homogeneous coordinates that represent points in the Euclidean plane and at infinity in a uniform way as triples. Projective transformations, which are invertible linear changes of homogeneous coordinates, are given by matrix multiplication. This lets computer graphics programs efficiently change the shape or the view of pictured objects and project them from three-dimensional virtual space to the two-dimensional viewing screen.

ALGEBRAIC GEOMETRY

Algebraic geometry extends the study of the geometric properties of solutions to polynomial equations beyond three dimensions. (Solutions in two and three dimensions are first covered in plane and solid analytic geometry, respectively.)

Algebraic geometry emerged from analytic geometry after 1850 when topology, complex analysis, and algebra were used to study algebraic curves. An algebraic curve C is the graph of an equation $f(x, y) = 0$, with points at infinity added, where $f(x, y)$ is a polynomial, in two complex variables, that cannot be factored. Curves are classified by a nonnegative integer—known as their genus, g—that can be calculated from their polynomial.

The equation $f(x, y) = 0$ determines y as a function of x at all but a finite number of points of C. Since x takes values in the complex numbers, which are two-dimensional over the real numbers, the curve C is two-dimensional over the real numbers near most of its points. C looks like a hollow sphere with g hollow handles attached and finitely many points pinched together—a sphere has genus 0, a torus has genus 1, and so forth. The Riemann-Roch theorem uses integrals along paths on C to characterize g analytically.

A birational transformation matches up the points on two curves via maps given in both directions by rational functions of the coordinates. Birational transformations preserve intrinsic properties of curves, such as their genus, but provide leeway for geometers to simplify and classify curves by eliminating singularities (problematic points).

An algebraic curve generalizes to a variety, which is the solution set of r polynomial equations in n complex variables. In general, the difference $n-r$ is the dimension of the variety—that is, the number of independent complex parameters near most points. For example, curves have (complex) dimension one and surfaces have (complex) dimension two. The French mathematician Alexandre Grothendieck revolutionized algebraic geometry in the 1950s by generalizing varieties to schemes and extending the Riemann-Roch theorem.

Arithmetic geometry combines algebraic geometry and number theory to study integer solutions of polynomial equations. It lies at the heart of the British mathematician Andrew Wiles's 1995 proof of Fermat's last theorem.

Projective Geometry

Projective geometry as a discipline originated with the French mathematician Girard Desargues (1591–1661) to deal with those properties of geometric figures that are not altered by projecting their image, or "shadow," onto

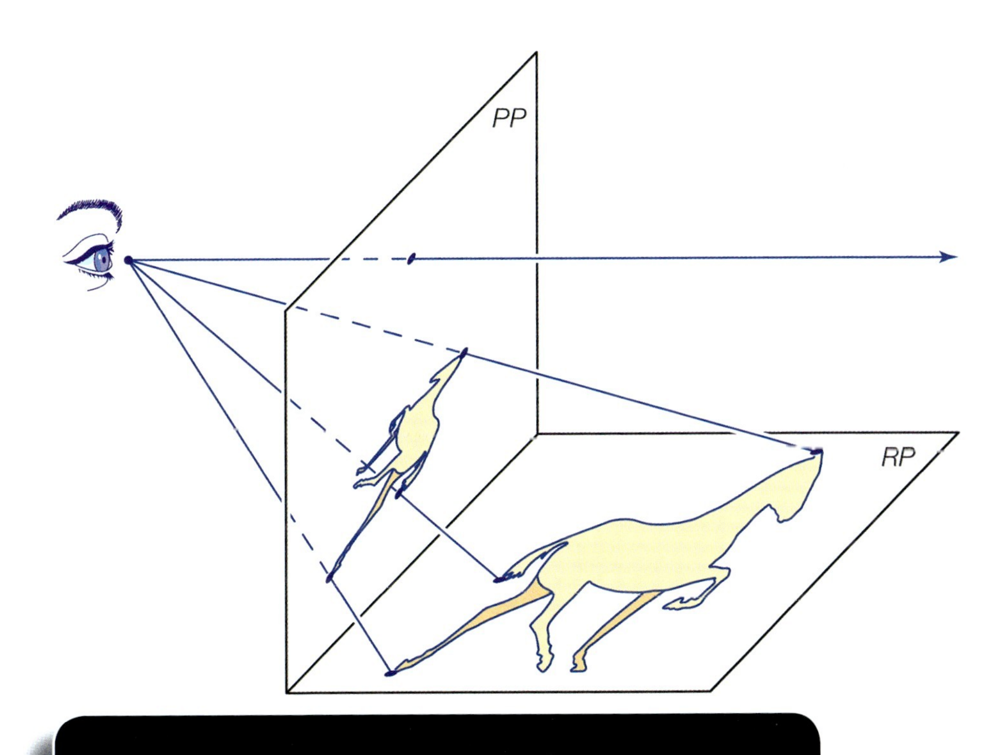

Projective drawing.

another surface. Common examples of projections are the shadows cast by opaque objects and motion pictures displayed on a screen.

Projective geometry has its origins in the early Italian Renaissance, particularly in the architectural drawings of Filippo Brunelleschi (1377–1446) and Leon Battista Alberti (1404–72), who invented the method of perspective drawing. By this method, the eye of the painter is connected to points on the landscape (the horizontal reality plane, *RP*) by so-called sight lines. The intersection of these sight lines with the vertical picture plane (*PP*) generates the drawing. Thus, the reality plane is projected onto the picture plane, hence the name projective geometry.

Although some isolated properties concerning projections were known in antiquity, particularly in the study of optics, it was not until the 17th century that mathematicians returned to the subject. The French mathematicians Girard Desargues (1591–1661) and Blaise Pascal (1623–62) took the first significant steps by examining what properties of figures were preserved (or invariant) under perspective mappings. The subject's real importance, however, became clear only after 1800 in the works of several other French mathematicians, notably Jean-Victor Poncelet (1788–1867). In general, by ignoring geometric measurements such as distances and angles, projective geometry enables a clearer understanding of some more generic properties of geometric objects. Such insights have since been incorporated in many more advanced areas of mathematics.

Parallel Lines and the Projection of Infinity

A theorem from Euclid's *Elements* (c. 300 BCE) states that if a line is drawn through a triangle such that it is parallel to one side, then the line will divide the other two sides proportionately; that is, the ratio of segments on each side will be equal. This is known as the proportional segments theorem, or the fundamental theorem of similarity, and for triangle ABC, with line segment DE parallel to side AB, the theorem corresponds to the mathematical expression $CD/DA = CE/EB$.

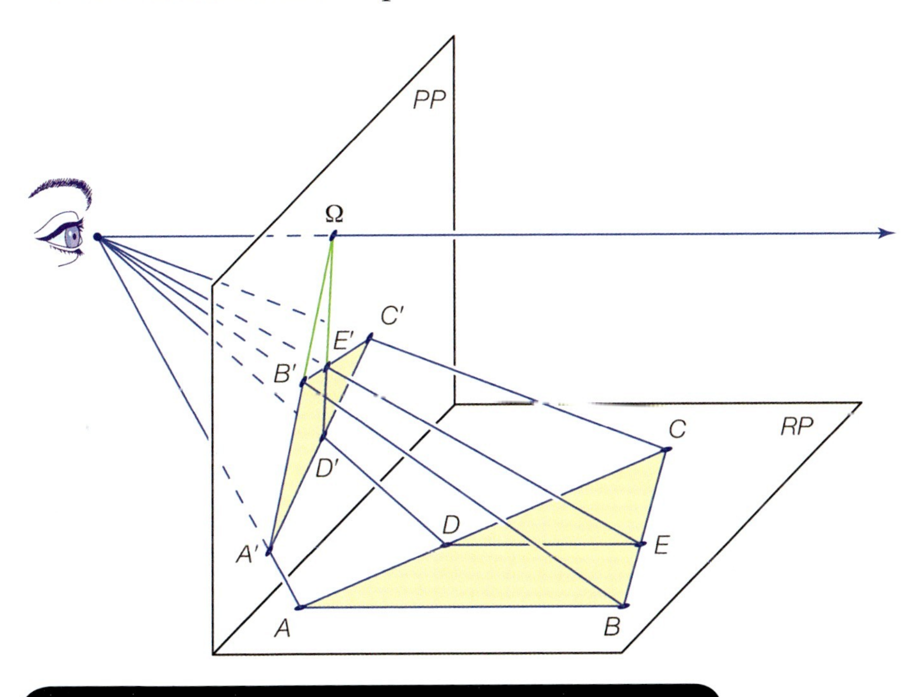

Projective version of the fundamental theorem of similarity.

Now consider the effect produced by projecting these line segments onto another plane. The first thing to note is that the projected line segments $A'B'$ and $D'E'$ are not parallel; i.e., angles are not preserved. From the point of view of the projection, the parallel lines AB and DE appear to converge at the horizon, or at infinity, whose projection in the picture plane is labeled Ω. (It was Desargues who first introduced a single point at infinity to represent the projected intersection of parallel lines. Furthermore, he collected all the points along the horizon in one line at infinity.) With the introduction of Ω, the projected figure corresponds to a theorem discovered by Menelaus of Alexandria in the 1st century CE:

$$C'D'/D'A' = C'E'/E'B' \cdot \Omega B'/\Omega A'.$$

Since the factor $\Omega B'/\Omega A'$ corrects for the projective distortion in lengths, Menelaus's theorem can be seen as a projective variant of the proportional segments theorem.

PROJECTIVE INVARIANTS

With Desargues's provision of infinitely distant points for parallels, the reality plane and the projective plane are essentially interchangeable—that is, ignoring distances and directions (angles), which are not preserved in the projection, other properties are preserved, however. For instance, two different points have a unique connecting line, and two different lines have a unique point of

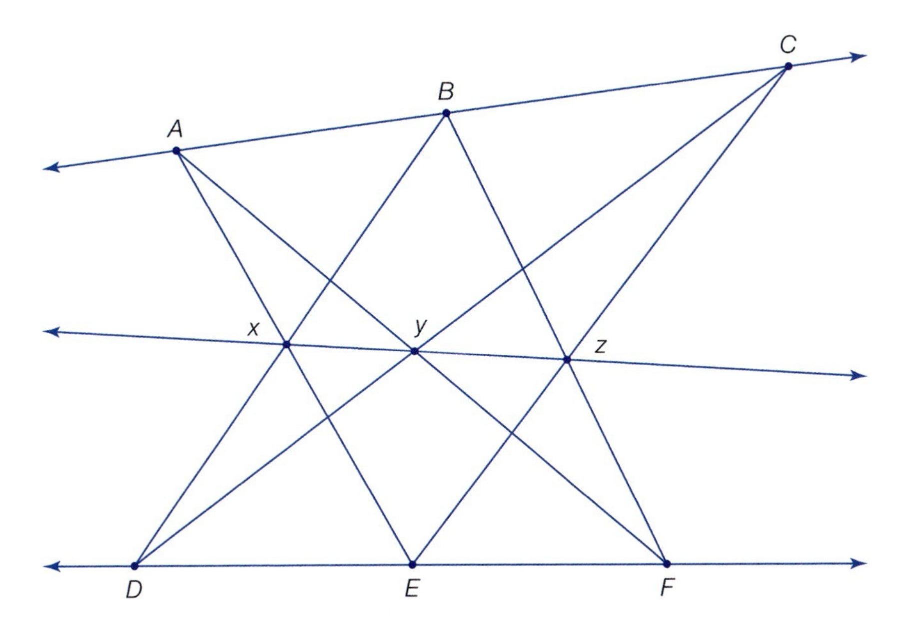

Pappus's projective theorem.

intersection. Although almost nothing else seems to be invariant under projective mappings, one should note that lines are mapped onto lines. This means that if three points are collinear (share a common line), then the same will be true for their projections. Thus, collinearity is another invariant property. Similarly, if three lines meet in a common point, so will their projections.

The following theorem is of fundamental importance for projective geometry. In its first variant, by Pappus of Alexandria (fl. 320 CE), it only uses collinearity:

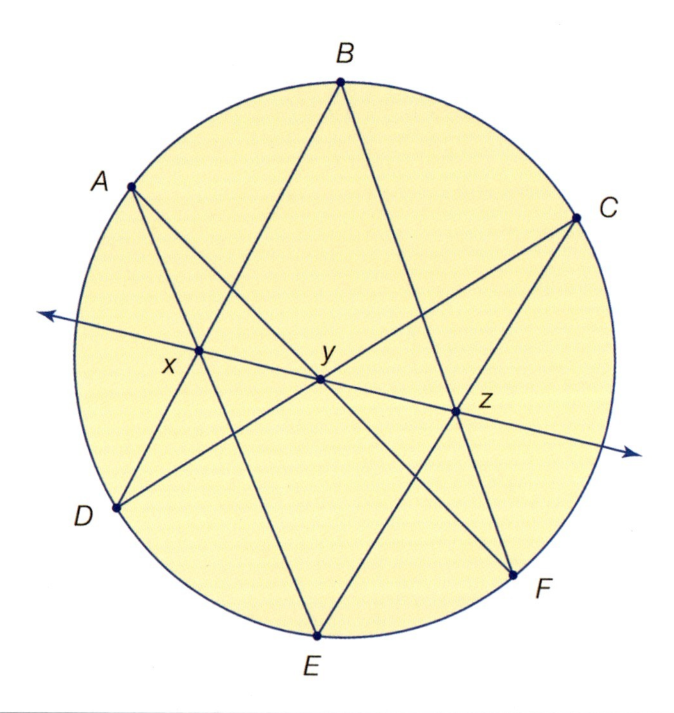

Pascal's projective theorem.

Let the distinct points A, B, C and D, E, F be on two different lines. Then the three intersection points—x of AE and BD, y of AF and CD, and z of BF and CE—are collinear.

The second variant, by Pascal, uses certain properties of circles:

If the distinct points A, B, C, D, E, and F are on one circle, then the three intersection points x, y, and z (defined as above) are collinear.

There is one more important invariant under projective mappings, known as the cross ratio. Given four distinct collinear points *A, B, C,* and *D,* the cross ratio is defined as

$$\text{CRat}(A, B, C, D) = AC/BC \cdot BD/AD.$$

It may also be written as the quotient of two ratios:

$$\text{CRat}(A, B, C, D) = AC/BC : AD/BD.$$

The latter formulation reveals the cross ratio as a ratio of ratios of distances. And while neither distance nor the ratio of distance is preserved under projection, Pappus first proved the startling fact that the cross ratio was invariant—that is,

$$\text{CRat}(A, B, C, D) = \text{CRat}(A', B', C', D').$$

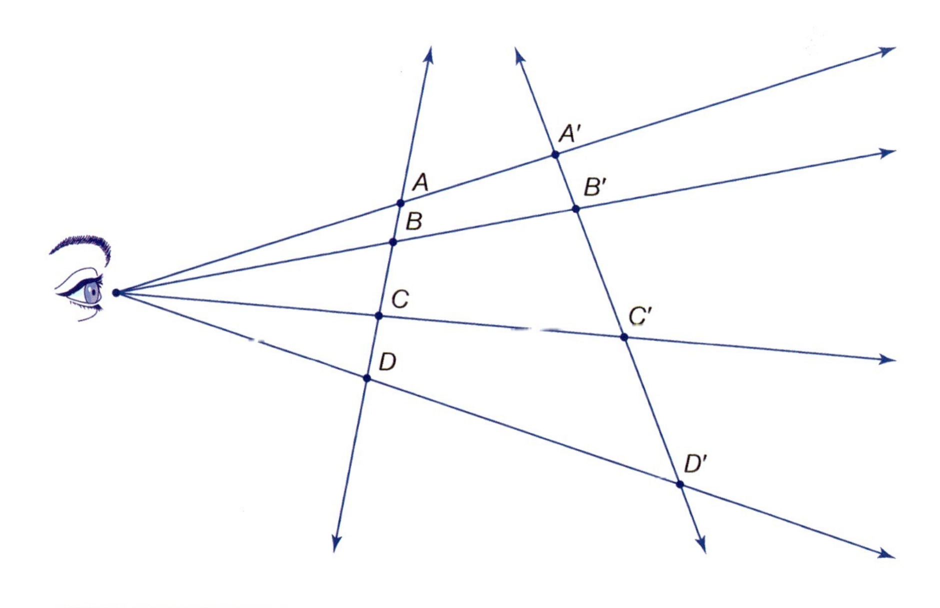

Cross ratio.

However, this result remained a mere curiosity until its real significance became gradually clear in the 19th century as mappings became more and more important for transforming problems from one mathematical domain to another.

PROJECTIVE CONIC SECTIONS

Conic sections can be regarded as plane sections of a right circular cone. By regarding a plane perpendicular to the cone's axis as the reality plane (*RP*), a "cutting" plane as the picture plane (*PP*), and the cone's apex as the projective "eye," each conic section can be seen to correspond to a projective image of a circle. Depending on the orientation of the cutting plane, the image of the circle will be a circle, an ellipse, a parabola, or a hyperbola.

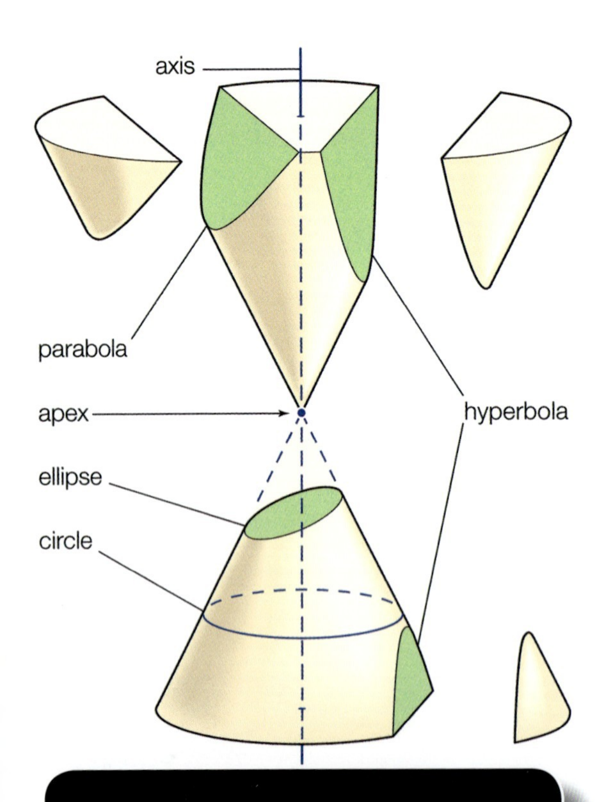

The conic sections result from intersecting a plane with a double cone.

A plane Ω passing through the apex and parallel to *PP* defines the line at infinity in the projective plane *PP*. The situation of Ω relative to *RP* determines the conic section in *PP*: If Ω intersects *RP* outside the base circle (the circle formed by the intersection of the cone and *RP*), the image of the circle will be an ellipse. If Ω is tangent to the base circle (in effect, tangent to the cone), the image will be a parabola. If Ω intersects the base circle (thus, cutting the circle in two), a hyperbola will result.

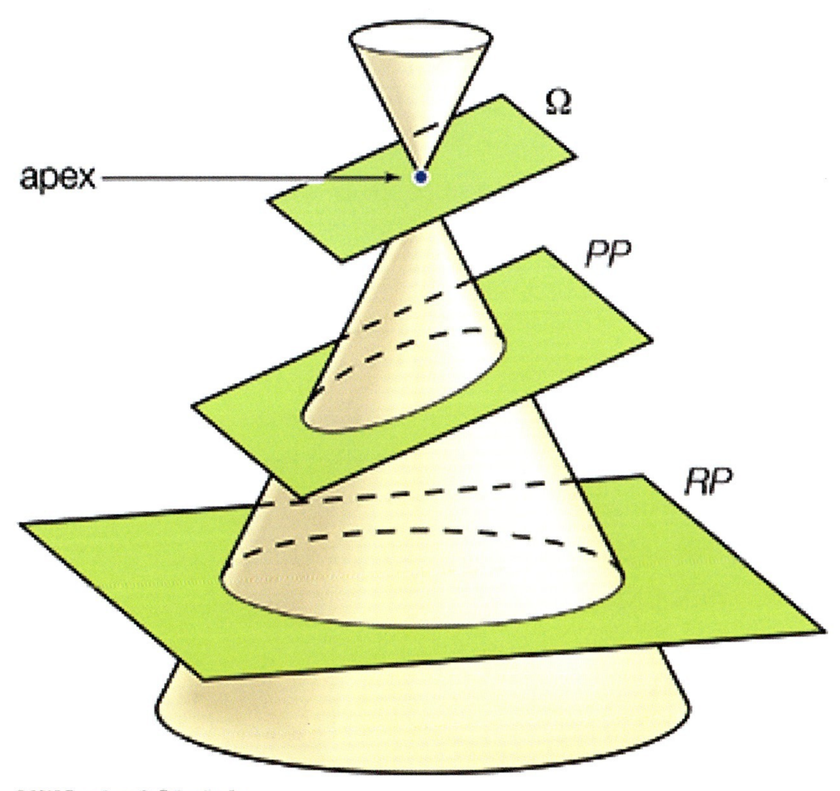

The conic sections can be generated by projecting the circle formed by the intersection of a cone with a plane perpendicular to the cone's central axis.

Pascal's theorem, quoted above, also follows easily for any conic section from its special case for the circle. Start by selecting six points on a conic section and project them back onto the base circle. As given earlier, the three relevant intersection points for six points on the circle will be collinear. Now project all nine points back to the conic section. Since collinear points (the three intersection points from the circle) are mapped onto collinear points, the theorem holds for any conic section. In this way the projective point of view unites the three different types of conics.

Similarly, more complicated curves and surfaces in higher-dimensional spaces can be unified through projections. For example, Isaac Newton (1643–1727) showed that all plane curves defined by polynomials in x and y of degree 3 (the highest power of the variables is 3) can be obtained as projective images of just five types of polynomials.

DIFFERENTIAL GEOMETRY

The German mathematician Carl Friedrich Gauss (1777–1855), in connection with practical problems of surveying and geodesy, initiated the field of differential geometry. Using differential calculus, he characterized the intrinsic properties of curves and surfaces. For instance, he showed that the intrinsic curvature of a cylinder is the same as that of a plane, as can be seen by cutting a cylinder along its axis and flattening, but not the same as that of a sphere, which cannot be flattened without distortion.

The discipline owes its name to its use of ideas and techniques from differential calculus, although the modern subject often uses algebraic and purely geometric techniques instead. Although basic definitions, notations, and analytic descriptions vary widely, the following geometric questions prevail: How does one measure the curvature of a curve within a surface (intrinsic) versus within the encompassing space (extrinsic)? How can the curvature of a surface be measured? What is the shortest path within a surface between two points on the surface? How is the shortest path on a surface related to the concept of a straight line?

While curves had been studied since antiquity, the discovery of calculus in the 17th century opened up the study of more complicated plane curves—such as those produced by the French mathematician René Descartes (1596–1650) with his "compass." In particular, integral calculus led to general solutions of the ancient problems of finding the arc length of plane curves and the area of plane figures. This in turn opened the stage to the investigation of curves and surfaces in space—an investigation that was the start of differential geometry.

Some of the fundamental ideas of differential geometry can be illustrated by the strake, a spiraling strip often designed by engineers to give structural support to large metal cylinders such as smokestacks. A strake can be formed by cutting an annular strip (the region between

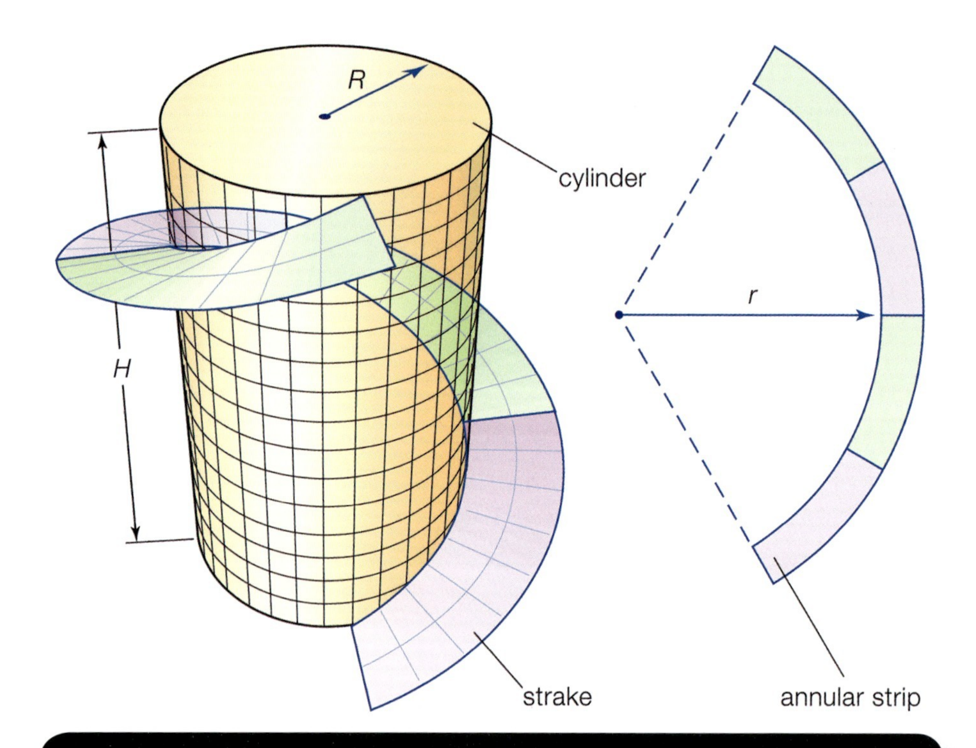

An annular strip can be cut and bent into a helical strake that follows approximately the contour of a cylinder.

two concentric circles) from a flat sheet of steel and then bending it into a helix that spirals around the cylinder. What should the radius r of the annulus be to produce the best fit? Differential geometry supplies the solution to this problem by defining a precise measurement for the curvature of a curve. Then r can be adjusted until the curvature of the inside edge of the annulus matches the curvature of the helix.

An important question remains: Can the annular strip be bent, without stretching, so that it forms a strake around the cylinder? In particular, this means that distances measured along the surface (intrinsic) are unchanged. Two surfaces are said to be isometric if one can be bent (or transformed) into the other without changing intrinsic distances. (For example, because a sheet of paper can be rolled into a tube without stretching, the sheet and tube are "locally" isometric—only locally because new, and possibly shorter, routes are created by connecting the two edges of the paper.) Thus, the second question becomes: Are the annular strip and the strake isometric? To answer this and similar questions, differential geometry developed the notion of the curvature of a surface.

CURVATURE OF CURVES

Although mathematicians from antiquity had described some curves as curving more than others and straight lines as not curving at all, it was the German mathematician Gottfried Leibniz who, in 1686, first defined the curvature of a curve at each point in terms of the circle that best approximates the curve at that point. Leibniz named his approximating circle the osculating circle, from the Latin *osculare* ("to kiss"). He then defined the curvature of the curve (and the circle) as $^1/_r$, where r is the radius of the osculating circle. As a curve becomes straighter, a circle with a larger radius must be used to approximate it, and so the resulting curvature decreases. In the limit, a straight

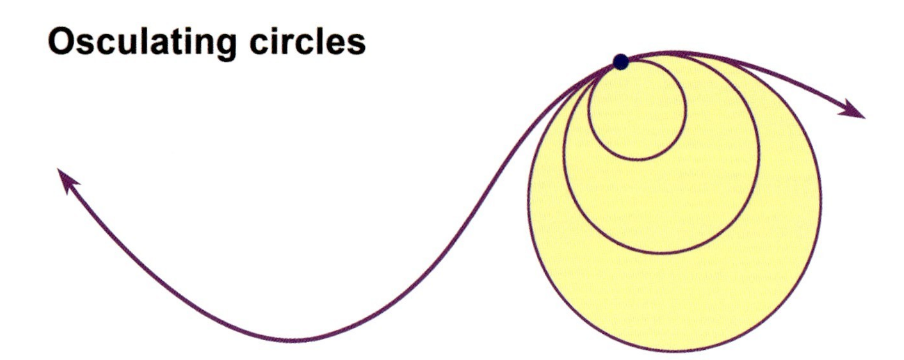

The curvature at each point of a line is defined to be $1/r$, where r is the radius of the osculating circle that best approximates the line at the given point.

line is said to be equivalent to a circle of infinite radius and its curvature defined as zero everywhere. The only curves in ordinary Euclidean space with constant curvature are straight lines, circles, and helices. In practice, curvature is found with a formula that gives the rate of change, or derivative, of the tangent to the curve as one moves along the curve. This formula was discovered by Isaac Newton and Leibniz for plane curves in the 17th century and by the Swiss mathematician Leonhard Euler for curves in space in the 18th century. (Note that the derivative of the tangent to the curve is not the same as the second derivative studied in calculus, which is the rate of change of the tangent to the curve as one moves along the x-axis.)

With these definitions in place, it is now possible to compute the ideal inner radius r of the annular strip that goes into making the strake shown in the figure. The annular strip's inner curvature $^1/_r$ must equal the curvature of the helix on the cylinder. If R is the radius of the cylinder and H is the height of one turn of the helix, then the curvature of the helix is $4\pi^2R/[H^2 + (2\pi R)^2]$. For example, if $R = 1$ metre and $H = 10$ metres, then $r = 3.533$ metres.

CURVATURE OF SURFACES

To measure the curvature of a surface at a point, Euler, in 1760, looked at cross sections of the surface made by planes that contain the line perpendicular (or "normal") to the surface at the point. Euler called the curvatures

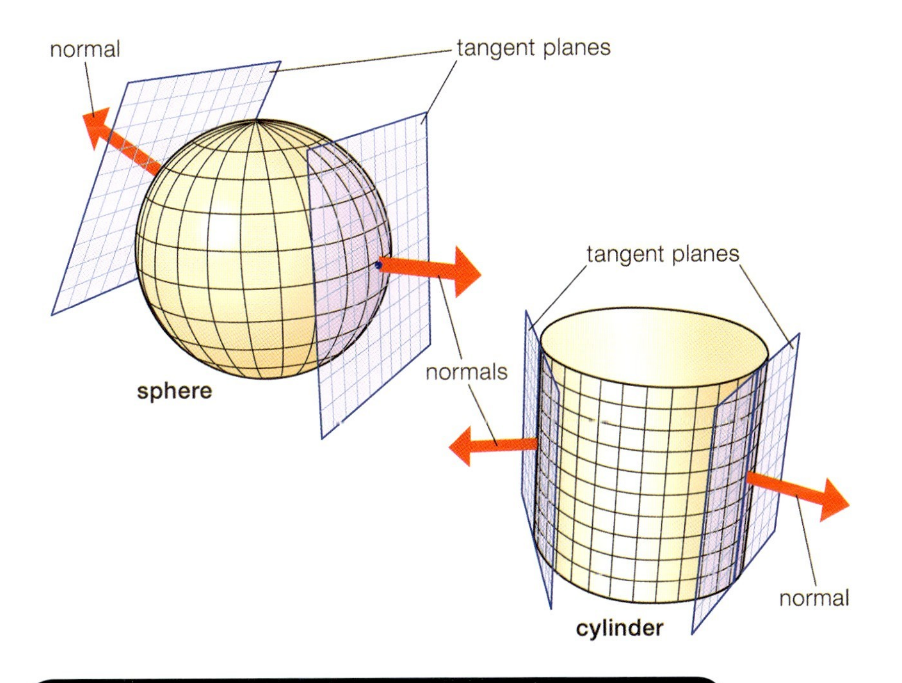

The normal, or perpendicular, at each point of a surface defines the corresponding tangent plane, and vice versa.

of these cross sections the normal curvatures of the surface at the point. For example, on a right cylinder of radius *r*, the vertical cross sections are straight lines and thus have zero curvature; the horizontal cross sections are circles, which have curvature $^1/_r$. The normal curvatures at a point on a surface are generally different in different directions. The maximum and minimum normal curvatures at a point on a surface are called the principal (normal) curvatures, and the directions in which these normal curvatures occur are called the principal directions. Euler proved that for most surfaces where the normal curvatures are not constant (for example, the cylinder), these principal directions are perpendicular to each other. (Note that on a sphere all the normal curvatures are the same and thus all are principal curvatures.) These principal normal curvatures are a measure of how "curvy" the surface is.

The theory of surfaces and principal normal curvatures was extensively developed by French geometers led by Gaspard Monge (1746–1818). It was in an 1827 paper, however, that the German mathematician Carl Friedrich Gauss made the big breakthrough that allowed differential geometry to answer the question raised above of whether the annular strip is isometric to the strake. The Gaussian curvature of a surface at a point is defined as the product of the two principal normal curvatures; it is said to be positive if the principal normal curvatures curve in the same direction and negative if they curve in opposite directions. Normal curvatures for a plane surface are all zero, and thus the Gaussian curvature of a plane is zero. For a cylinder of radius *r*, the minimum normal curvature is zero (along the vertical straight lines), and the maximum is $^1/_r$ (along the horizontal circles). Thus, the Gaussian curvature of a cylinder is also zero.

If the cylinder is cut along one of the vertical straight lines, the resulting surface can be flattened (without stretching) onto a rectangle. In differential geometry, it is said that the plane and cylinder are locally isometric. These are special cases of two important theorems:

> *Gauss's "Remarkable Theorem"* (1827). If two smooth surfaces are isometric, then the two surfaces have the same Gaussian curvature at corresponding points. (Athough defined extrinsically, Gaussian curvature is an intrinsic notion.)

> *Minding's theorem* (1839). Two smooth ("cornerless") surfaces with the same constant Gaussian curvature are locally isometric.

As corollaries to these theorems:

> A surface with constant positive Gaussian curvature *c* has locally the same intrinsic geometry as a sphere of radius $\sqrt{^1/_c}$. (This is because a

sphere of radius r has Gaussian curvature $1/r^2$).

A surface with constant zero Gaussian curvature has locally the same intrinsic geometry as a plane. (Such surfaces are called developable).

A surface with constant negative Gaussian curvature c has locally the same intrinsic geometry as a hyperbolic plane.

The Gaussian curvature of an annular strip (being in the plane) is constantly zero. So to answer whether or not the annular strip is isometric to the strake, one needs only to check whether a strake has constant zero Gaussian curvature. The Gaussian curvature of a strake is actually negative, hence the annular strip must be stretched—although this can be minimized by narrowing the shapes.

SHORTEST PATHS ON A SURFACE

From an outside, or extrinsic, perspective, no curve on a sphere is straight. Nevertheless, the great circles are intrinsically straight—an ant crawling along a great circle does not turn or curve with respect to the surface. About 1830 the Estonian mathematician Ferdinand Minding defined a curve on a surface to be a geodesic if it is intrinsically straight—that is, if there is no identifiable curvature from within the surface.

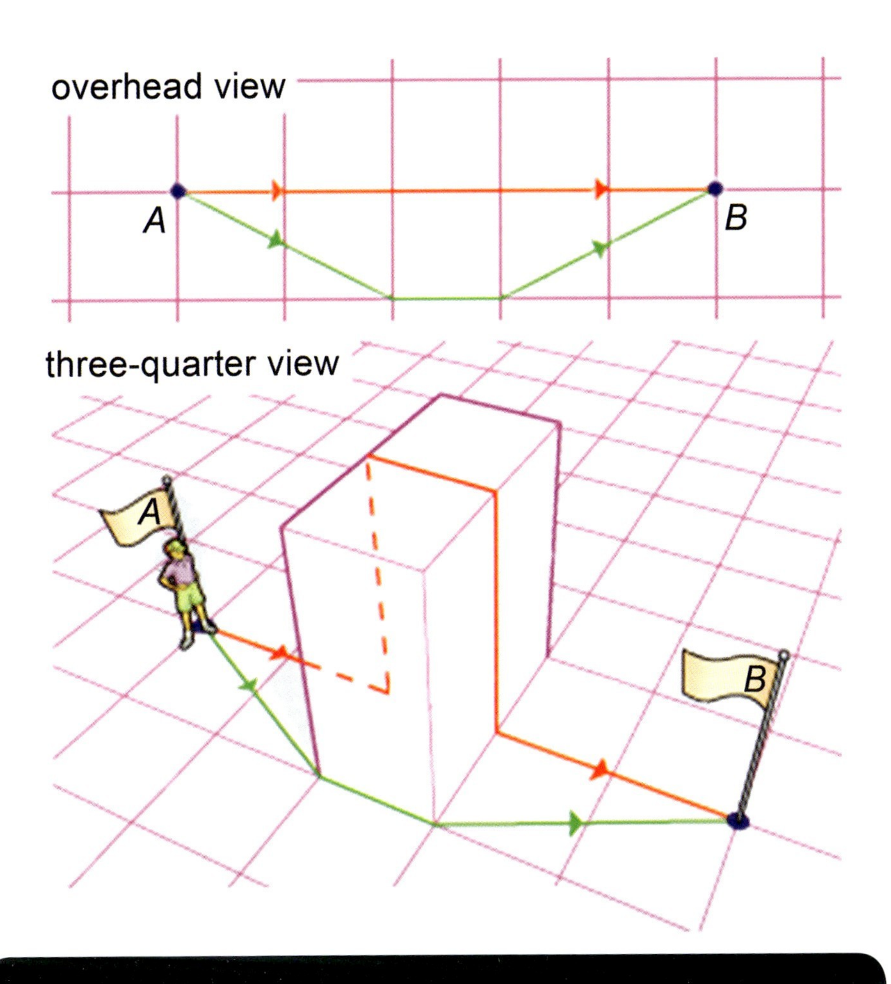

The red path from *A* to *B* that rises over the elevation is intrinsically straight (as viewed from within the surface).

A major task of differential geometry is to determine the geodesics on a surface. The great circles are the geodesics on a sphere.

A great circle arc that is longer than a half circle is intrinsically straight on the sphere, but it is not the shortest distance between its endpoints. On the other hand, the shortest path in a surface is not always straight. An important theorem is:

On a surface which is complete (every geodesic can be extended indefinitely) and smooth, every shortest curve is intrinsically straight and every intrinsically straight curve is the shortest curve between nearby points.

NON-EUCLIDEAN GEOMETRY

Beginning in the 19th century, various mathematicians substituted alternatives to Euclid's parallel postulate, which, in its modern form, reads, "given a line and a point not on the line, it is possible to draw exactly one line through the given point parallel to the line." They hoped to show that the alternatives were logically impossible. Instead, they discovered that consistent non-Euclidean geometries exist.

Although non-Euclidean geometry is frequently used to refer only to hyperbolic geometry, common usage includes those few geometries (hyperbolic and spherical) that differ from but are very close to Euclidean geometry.

COMPARISON OF EUCLIDEAN, SPHERICAL, AND HYPERBOLIC GEOMETRIES	
Given a line and a point not on the line, there exist(s) through the given point and parallel to the given line	
exactly one line	Euclidean
no lines	spherical
infinitely many lines	hyperbolic
Euclid's fifth postulate is	
true	Euclidean
false	spherical
false	hyperbolic
The sum of the interior angles of a triangle 180 degrees	
=	Euclidean
>	spherical
<	hyperbolic

The non-Euclidean geometries developed along two different historical threads. The first thread started with the search to understand the movement of stars and planets in the apparently hemispherical sky. For example, Euclid (fl. c. 300 BCE) wrote about spherical geometry in his astronomical work *Phaenomena*. In addition to looking to the heavens, the ancients attempted to understand the shape of the Earth and to use this understanding to solve problems in navigation over long distances (and later for large-scale surveying). These activities are aspects of spherical geometry.

The second thread started with the fifth ("parallel") postulate in Euclid's *Elements*:

If a straight line falling on two straight lines makes the interior angles on the same side less than two right angles, the two straight lines, if produced indefinitely, will meet on that side on which the angles are less than the two right angles.

For 2,000 years following Euclid, mathematicians attempted either to prove the postulate as a theorem (based on the other postulates) or to modify it in various ways. These attempts culminated when the Russian Nikolay Lobachevsky (1829) and the Hungarian János Bolyai (1831) independently published a description of a geometry that, except for the parallel postulate, satisfied all of Euclid's postulates and common notions. This geometry is called hyperbolic geometry.

Spherical Geometry

From early times, people noticed that the shortest distance between two points on Earth were great circle routes. For example, the Greek astronomer Ptolemy wrote in *Geography* (c. 150 CE):

It has been demonstrated by mathematics that the surface of the land and water is in its entirety a sphere . . . and that any plane which passes through the centre makes at its surface, that is, at the surface of the Earth and of the sky, great circles.

Great circles are the "straight lines" of spherical geometry. This is a consequence of the properties of a sphere, in which the shortest distances on the surface are great circle routes. Such curves are said to be "intrinsically" straight. (Note, however, that intrinsically straight and shortest are not necessarily identical.) Three intersecting great circle arcs form a spherical triangle; while a spherical triangle must be distorted to fit on another sphere with a different radius, the difference is only one of scale. In differential geometry, spherical geometry is described as the geometry of a surface with constant positive curvature.

There are many ways of projecting a portion of a sphere, such as the surface of the Earth, onto a plane. These are known as maps or charts and they must necessarily distort distances and either area or angles. Cartographers' need for various qualities in map projections gave an early impetus to the study of spherical geometry.

Elliptic geometry is the term used to indicate an axiomatic formalization of spherical geometry in which each pair of antipodal points is treated as a single point. An intrinsic analytic view of spherical geometry was developed in the 19th century by the German mathematician Bernhard Riemann; usually called the Riemann sphere, it is studied in university courses on complex analysis. Some texts call this (and therefore spherical geometry) Riemannian geometry, but this term more correctly applies to a part of differential geometry that gives a way of intrinsically describing any surface.

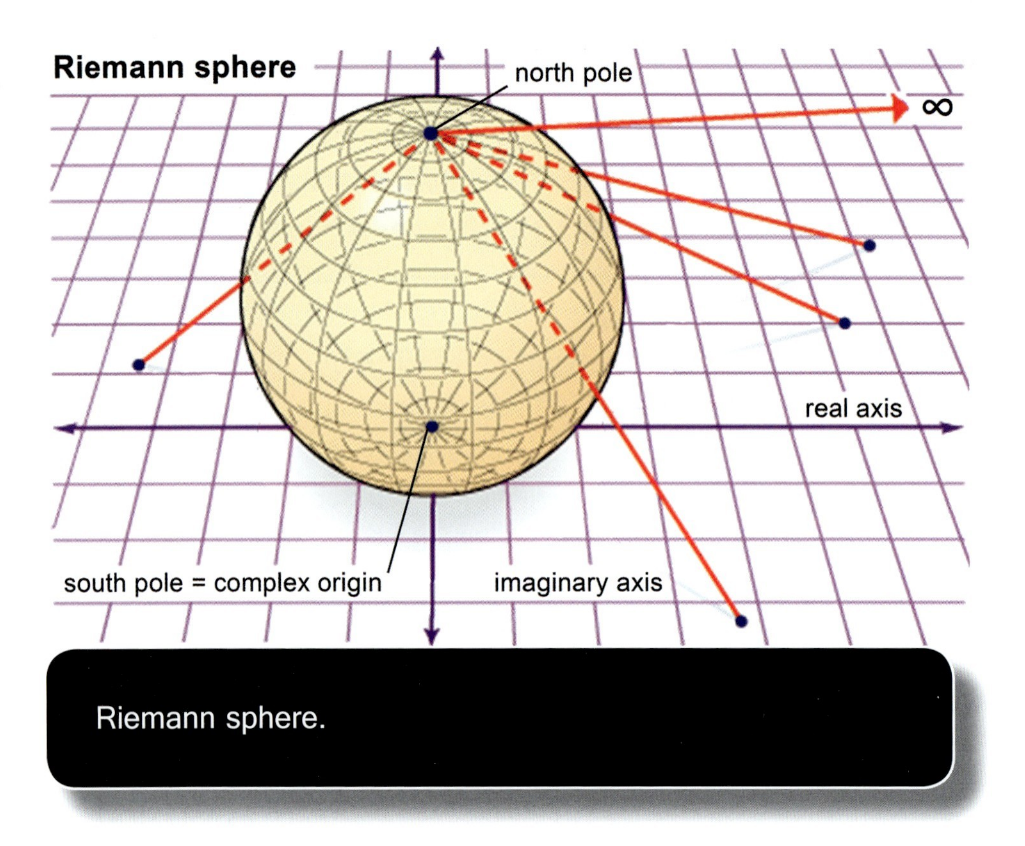

Riemann sphere.

Hyperbolic Geometry

The first description of hyperbolic geometry was given in the context of Euclid's postulates, and it was soon proved that all hyperbolic geometries differ only in scale (in the same sense that spheres only differ in size). In the mid19th century it was shown that hyperbolic surfaces must have constant negative curvature. However, this still left open the question of whether any surface with hyperbolic geometry actually exists.

In 1868 the Italian mathematician Eugenio Beltrami described a surface, called the pseudosphere, that has constant negative curvature. However, the pseudosphere is not a complete model for hyperbolic geometry, because intrinsically straight lines on the pseudosphere may intersect themselves and cannot be continued past the bounding circle (neither of which is true in hyperbolic geometry). In 1901 the German mathematician David Hilbert proved that it is impossible to define a complete hyperbolic surface using real analytic functions (essentially, functions that can be expressed in terms of ordinary formulas). In those days, a surface always meant one defined by real analytic functions, and so the search was abandoned. However, in 1955 the Dutch mathematician Nicolaas Kuiper proved the existence of a complete hyperbolic surface, and in the 1970s the American mathematician William Thurston described the construction of a knot that had a hyperbolic surface. This surface could be crocheted.

In the 19th century, mathematicians developed three models of hyperbolic geometry that can now be interpreted as projections (or maps) of the hyperbolic surface. Although these models all suffer from some distortion—similar to the way that flat maps distort the spherical Earth—they are useful individually and in combination as aides to understand hyperbolic geometry. In 1869–71 Beltrami and the German mathematician Felix Klein developed the first complete model of hyperbolic geometry (and first called the geometry "hyperbolic"). In the Klein-Beltrami model, the hyperbolic surface is mapped to the interior of a circle, with geodesics in the hyperbolic surface corresponding to chords in the circle. Thus, the Klein-Beltrami model preserves "straightness" but at the cost of distorting angles. About 1880 the French mathematician Henri Poincaré developed two more models. In the Poincaré disk model, the hyperbolic surface is mapped to the interior of a circular disk, with hyperbolic geodesics mapping to circular arcs (or diameters) in the disk that meet the bounding circle at right angles. In the Poincaré upper half-plane model, the hyperbolic surface is mapped onto the half-plane above the x-axis, with hyperbolic geodesics mapped to semicircles (or vertical rays) that meet the x-axis at right angles. Both Poincaré models distort distances while preserving angles as measured by tangent lines.

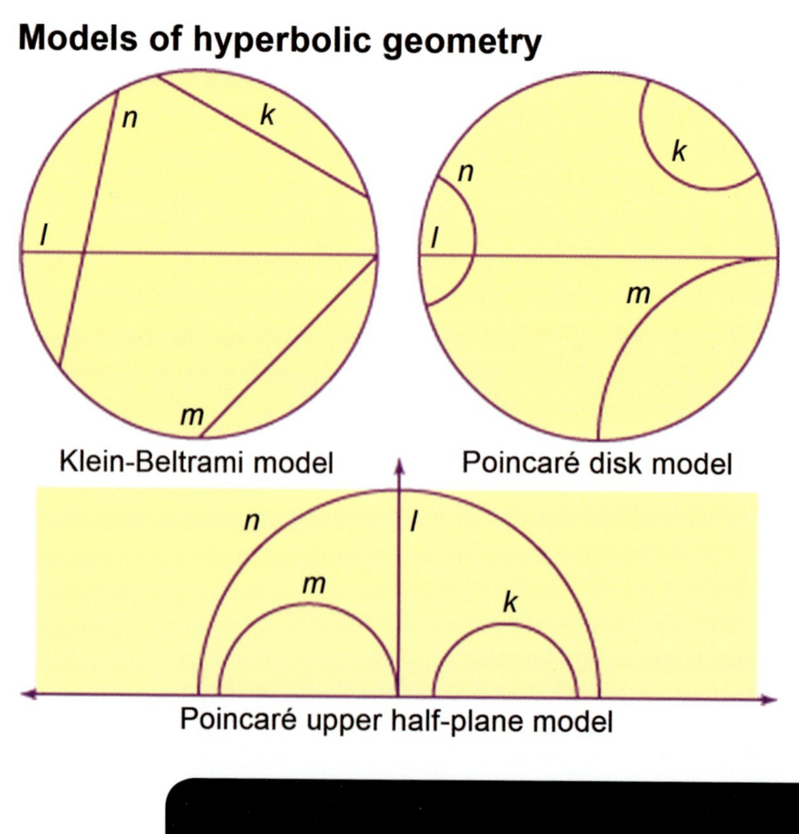

Poincaré disk model.

TOPOLOGY

Topology, the youngest and most sophisticated branch of geometry, focuses on the properties of geometric objects that remain unchanged upon continuous deformation—shrinking, stretching, and folding, while disallowing tearing apart or gluing together parts. The continuous development of topology dates from 1911, when the Dutch mathematician L.E.J. Brouwer (1881–1966) introduced methods generally applicable to the topic.

Topology, while similar to geometry, differs from geometry in that geometrically equivalent objects often share numerically measured quantities, such as lengths or angles, while topologically equivalent objects resemble each other in a more qualitative sense.

The area of topology dealing with abstract objects is referred to as general, or point-set, topology. General topology overlaps with another important area of topology called algebraic topology. These areas of specialization form the two major subdisciplines of topology that developed during its relatively modern history.

SIMPLY CONNECTED

In some cases, the objects considered in topology are ordinary objects residing in three- (or lower-) dimensional space. For example, a simple loop in a plane and the boundary edge of a square in a plane are topologically equivalent, as may be observed by imagining the loop as

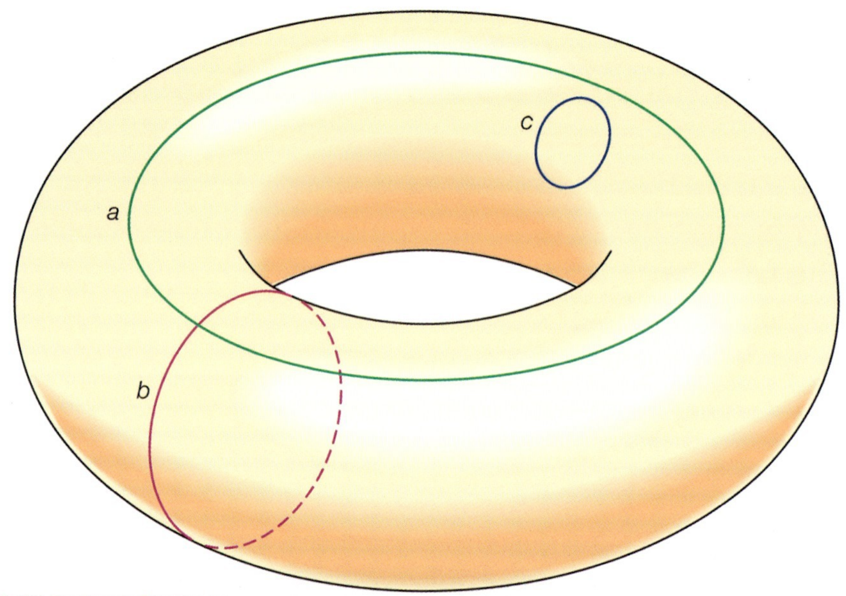

While the small loop *c* can be shrunk to a point without breaking the loop or the torus, loops *a* and *b* cannot because they encompass the torus's central hole.

a rubber band that can be stretched to fit tightly around the square. On the other hand, the surface of a sphere is not topologically equivalent to a torus, the surface of a solid doughnut ring. To see this, note that any small loop lying on a fixed sphere may be continuously shrunk, while being kept on the sphere, to any arbitrarily small diameter. An object possessing this property is said to be simply connected, and the property of being simply connected is indeed a property retained under a continuous deformation. However, some loops on a torus cannot be shrunk.

Many results of topology involve objects as simple as those mentioned above. The importance of topology as a branch of mathematics, however, arises from its more general consideration of objects contained in higher-dimensional spaces or even abstract objects that are sets of elements of a very general nature. To facilitate this generalization, the notion of topological equivalence must be clarified.

TOPOLOGICAL EQUIVALENCE

The motions associated with a continuous deformation from one object to another occur in the context of some surrounding space, called the ambient space of the deformation. When a continuous deformation from one object to another can be performed in a particular ambient space, the two objects are said to be isotopic with respect to that space. For example, consider an object that consists of a circle and an isolated point inside the circle. Let a second object consist of a circle and an isolated point outside the circle, but in the same plane as the circle. In a two-dimensional ambient space, these two objects cannot be continuously deformed into each other because it would require cutting the circles open to allow the isolated points to pass through. However, if three-dimensional space serves as the ambient space, a continuous deformation can be performed—simply lift the isolated point out of the plane and reinsert it on the other side of the circle to accomplish the task.

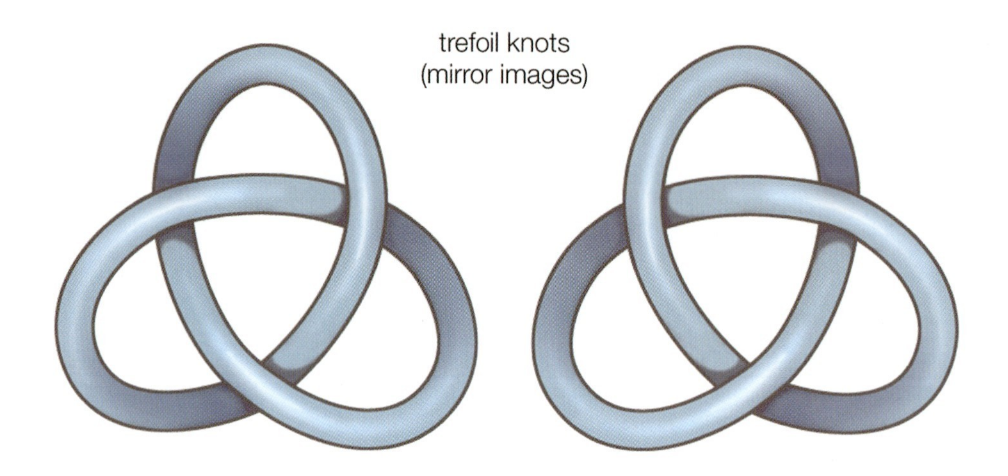

The trefoil knot is the only knot, other than its mirror image, that can be formed with exactly three crossings.

Thus, these two objects are isotopic with respect to three-dimensional space, but they are not isotopic with respect to two-dimensional space.

The notion of objects being isotopic with respect to a larger ambient space provides a definition of extrinsic topological equivalence, in the sense that the space in which the objects are embedded plays a role. The example above motivates some interesting and entertaining extensions. One might imagine a pebble trapped inside a spherical shell. In three-dimensional space the pebble cannot be removed without cutting a hole through the shell, but by adding an abstract fourth dimension it can be removed without any such surgery. Similarly, a closed loop of rope that is tied as a trefoil, or overhand, knot in three-dimensional space can be untied in an abstract four-dimensional space.

HOMEOMORPHISM

An intrinsic definition of topological equivalence (independent of any larger ambient space) involves a special type of function known as a homeomorphism. A function h is a homeomorphism, and objects x and y are said to be homeomorphic, if and only if the function satisfies the following conditions.

(1) h is a one-to-one correspondence between the elements of x and Y;

(2) h is continuous: nearby points of x are mapped to nearby points of y and distant points of x are mapped to distant points of Y—in other words, "neighbourhoods" are preserved;

(3) there exists a continuous inverse function h^{-1}: thus, $h^{-1}h(x) = x$ for all $x \in x$ and $hh^{-1}(y) = y$ for all $y \in Y$—in other words, there exists a function that "undoes" (is the inverse of) the homeomorphism, so that for any x in x or any y in y the original value can be restored by combining the two functions in the proper order.

The notion of two objects being homeomorphic provides the definition of intrinsic topological equivalence and is the generally accepted meaning of topological equivalence. Two objects that are isotopic in some ambient space must also be homeomorphic. Thus, extrinsic topological equivalence implies intrinsic topological equivalence.

TOPOLOGICAL STRUCTURE

In its most general setting, topology involves objects that are abstract sets of elements. To discuss properties such as continuity of functions between such abstract sets, some additional structure must be imposed on them.

TOPOLOGICAL SPACE

One of the most basic structural concepts in topology is to turn a set x into a topological space by specifying a collection of subsets T of X. Such a collection must satisfy three axioms: (1) the set x itself and the empty set are members of T, (2) the intersection of any finite number of sets in T is in T, and (3) the union of any collection of sets in T is in T. The sets in T are called open sets and T is called a topology on X. For example, the real number line becomes a topological space when its topology is specified as the collection of all possible unions of open intervals—such as $(-5, 2)$, $(1/2, \pi)$, $(0, \sqrt{2})$, …. (An analogous process produces a topology on a metric space.) Other examples of topologies on sets occur purely in terms of set theory. For example, the collection of all subsets of a set *x* is called the discrete topology on *X*, and the collection consisting only of the empty set and *x* itself forms the indiscrete, or trivial, topology on *X*. A given topological space gives rise to other related topological spaces. For example, a subset *A* of a topological space *x* inherits a topology, called the relative topology, from *x* when the open sets of *A* are taken to be the intersections of *A* with open sets of *X*. The tremendous variety of topological spaces provides a rich source of examples to motivate general theorems, as well as counterexamples to demonstrate false conjectures. Moreover, the generality of the axioms for a topological space permit mathematicians to view many sorts of mathematical structures, such as collections of functions in analysis, as topological spaces and thereby explain associated phenomena in new ways.

A topological space may also be defined by an alternative set of axioms involving closed sets, which are complements of open sets. In early consideration of topological ideas, especially for objects in *n*-dimensional Euclidean space, closed sets had arisen naturally in the investigation of convergence of infinite sequences. It is often convenient or useful to assume extra axioms for a topology in order to establish results that hold for a significant class of topological spaces but not for all topological spaces. One such axiom requires that two distinct points should belong to disjoint open sets. A topological space satisfying this axiom has come to be called a Hausdorff space.

CONTINUITY

An important attribute of general topological spaces is the ease of defining continuity of functions. A function f mapping a topological space x into a topological space y is defined to be continuous if, for each open set V of Y, the subset of x consisting of all points p for which $f(p)$ belongs to V is an open set of X. Another version of this definition is easier to visualize. A function f from a topological space x to a topological space y is continuous at $p \in x$ if, for any neighbourhood V of $f(p)$, there exists a neighbourhood U of p such that $f(U) \subseteq V$. These definitions provide important generalizations of the usual notion of continuity studied in analysis and also allow for a straightforward generalization of the notion of homeomorphism to the case of general topological spaces. Thus, for general topological spaces, invariant properties are those preserved by homeomorphisms.

ALGEBRAIC TOPOLOGY

The idea of associating algebraic objects or structures with topological spaces arose early in the history of topology. The basic incentive in this regard was to find topological invariants associated with different structures. The simplest example is the Euler characteristic, which is a number associated with a surface. In 1750 the Swiss mathematician Leonhard Euler proved the polyhedral formula $V - E + F = 2$, or Euler characteristic,

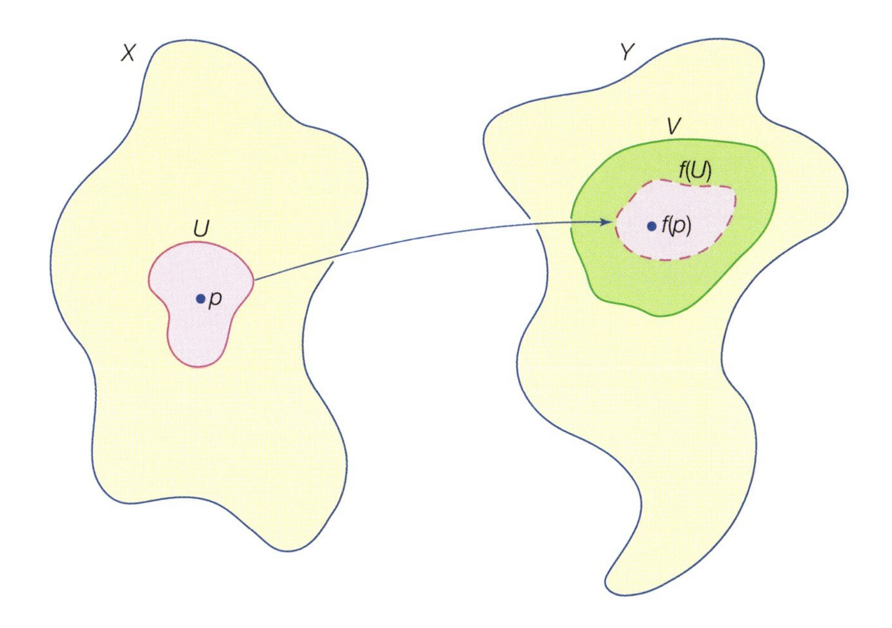

The topological concept of a continuous function.

which relates the numbers V and E of vertices and edges, respectively, of a network that divides the surface of a polyhedron (being topologically equivalent to a sphere) into F simply connected faces. This simple formula motivated many topological results once it was generalized to the analogous Euler-Poincaré characteristic $\chi = V - E + F = 2 - 2g$ for similar networks on the surface of a g-holed torus. Two homeomorphic surfaces will have the same Euler-Poincaré characteristic, and so two surfaces with different Euler-Poincaré characteristics cannot be topologically equivalent. However, the primary algebraic objects used in algebraic topology are more intricate and include such structures as abstract groups, vector spaces, and sequences of groups. Moreover, the language of algebraic topology has been enhanced by the introduction of category theory, in which very general mappings translate topological spaces and continuous functions between them to the associated algebraic objects and their natural mappings, which are called homomorphisms.

FUNDAMENTAL GROUP

A very basic algebraic structure called the fundamental group of a topological space was among the algebraic ideas studied by the French mathematician Henri Poincaré in the late 19th century. This group essentially consists of curves in the space that are combined by an operation arising in a geometric way. While this group was well understood even in the early days of algebraic topology for compact two-dimensional surfaces, some questions related to it still remain unanswered, especially for certain compact manifolds, which generalize surfaces to higher dimensions.

The most famous of these questions, called the Poincaré conjecture, asks if a compact three-dimensional manifold with trivial fundamental group is necessarily homeomorphic to the three-dimensional sphere (the set of points in four-dimensional space that are equidistant from the origin), as is known to be true for the two-dimensional case. Much research in algebraic topology has been related in some way to this conjecture since it was posed by Poincaré in 1904. One such research effort concerned a conjecture on the geometrization of three-dimensional manifolds that was posed in the 1970s by the American mathematician William Thurston. Thurston's conjecture implies the Poincaré conjecture, and in recognition of his work toward proving these conjectures, the Russian mathematician Grigori Perelman was awarded a Fields Medal at the 2006 International Congress of Mathematicians.

KNOT THEORY

A branch of algebraic topology that is involved in the study of three-dimensional manifolds is knot theory, the study of the ways in which knotted copies of a circle can be embedded in three-dimensional space. Knot theory, which dates back to the late 19th century, gained increased attention in the last two decades of the 20th century when its potential applications in physics, chemistry, and biomedical engineering were recognized.

Knots may be regarded as formed by interlacing and looping a piece of string in any fashion and then joining the ends. The first question that arises is whether such a curve is truly knotted or can simply be untangled; that is, whether or not one can deform it in space into a standard unknotted curve like a circle. The second question is whether, more generally, any two given curves represent different knots or are really the same knot in the sense that one can be continuously deformed into the other.

(Continued on the next page)

(Continued from the previous page)

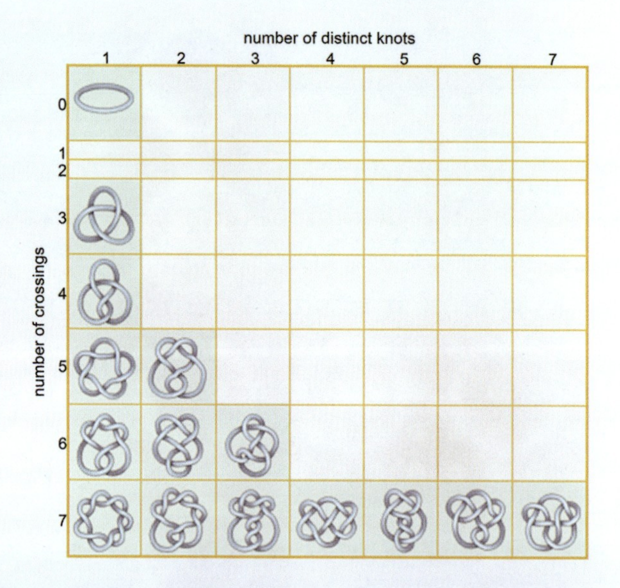

Knots are characterized by the number of times and the manner in which the strand crosses itself.

The basic tool for classifying knots consists of projecting each knot onto a plane—picture the shadow of the knot under a light—and counting the number of times the projection crosses itself, noting at each crossing which direction goes "over" and which goes "under." A measure of the knot's complexity is the least number of crossings that occur as the knot is moved around in all possible ways. The simplest possible true knot is the trefoil knot, or overhand knot, which has three such crossings. The order of this knot is therefore denoted as three. Even this simple knot has two configurations that cannot be deformed into each other, although they are mirror images. There are no knots with fewer crossings, and all others have at least four.

The number of distinguishable knots increases rapidly as the order increases. For example, there are almost 10,000 distinct knots with 13 crossings, and over a million with 16 crossings—the highest known by the end of the 20th century. Certain higher-order knots can be resolved into combinations, called products, of lower-order knots. For example, the square knot and the granny knot (sixth order knots) are products of two trefoils that are of the same or opposite chirality, or handedness. Knots that cannot be so resolved are called prime.

The first steps toward a mathematical theory of knots were taken about 1800 by the German mathematician Carl Friedrich Gauss. The origins of modern knot theory, however, stem from a suggestion by the Scottish mathematician-physicist William Thomson (Lord Kelvin) in 1869 that atoms might consist of knotted vortex tubes of the ether, with different elements corresponding to different knots. In response, a contemporary, the Scottish mathematician-physicist Peter Guthrie Tait, made the first systematic attempt to classify knots. Although Kelvin's theory was eventually rejected along with ether, knot theory continued to develop as a purely mathematical theory for about 100 years. Then a major breakthrough by the New Zealand mathematician Vaughan Jones in 1984, with the introduction of the Jones polynomials as new knot invariants, led the American mathematical physicist Edward Witten to discover a connection between knot theory and quantum field theory. (Both men were awarded Fields Medals in 1990 for their work.) In another direction, the American mathematician (and fellow Fields medalist) William Thurston made an important link between knot theory and hyperbolic geometry, with possible ramifications in cosmology. Other applications of knot theory have been made in biology, chemistry, and mathematical physics.

The fundamental group is the first of what are known as the homotopy groups of a topological space. These groups, as well as another class of groups called homology groups, are actually invariant under mappings called homotopy retracts, which include homeomorphisms. Homotopy theory and homology theory are among the many specializations within algebraic topology.

GRAPH THEORY

The history of graph theory may be traced to 1735, when the Swiss mathematician Leonhard Euler solved the Königsberg bridge problem. The Königsberg bridge problem was an old puzzle concerning the possibility of finding a path over every one of seven bridges that span a forked river flowing past an island—but without cross-

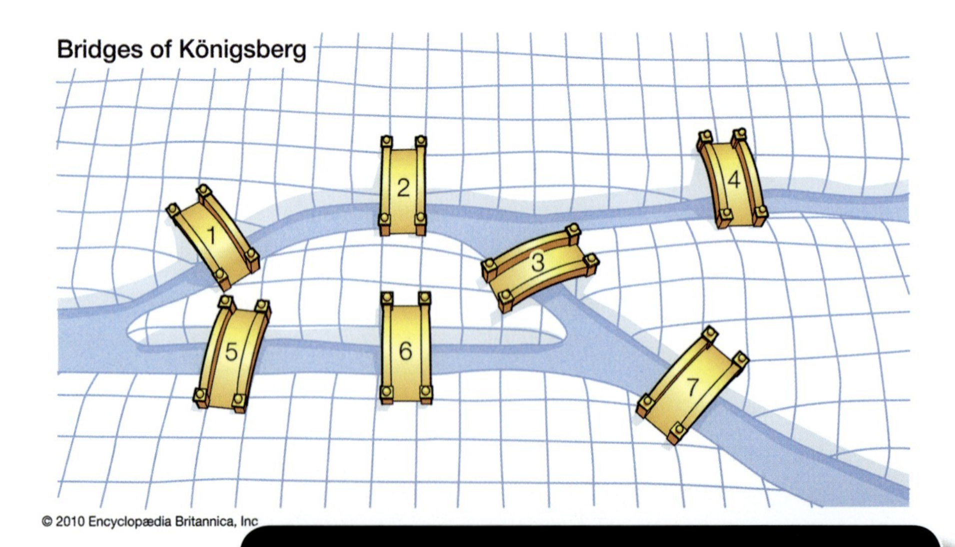

Euler was intrigued by the question of whether a route existed that would traverse each of the seven bridges exactly once.

ing any bridge twice. Euler argued that no such path exists. His proof involved only references to the physical arrangement of the bridges, but essentially he proved the first theorem in graph theory. Translated into the terminology of modern graph theory (introduced much later), the theorem could be restated as follows: If there is a path along edges of a multigraph that traverses each edge once and only once, then there exist at most two vertices of odd degree; furthermore, if the path begins and ends at the same vertex, then no vertices will have odd degree.

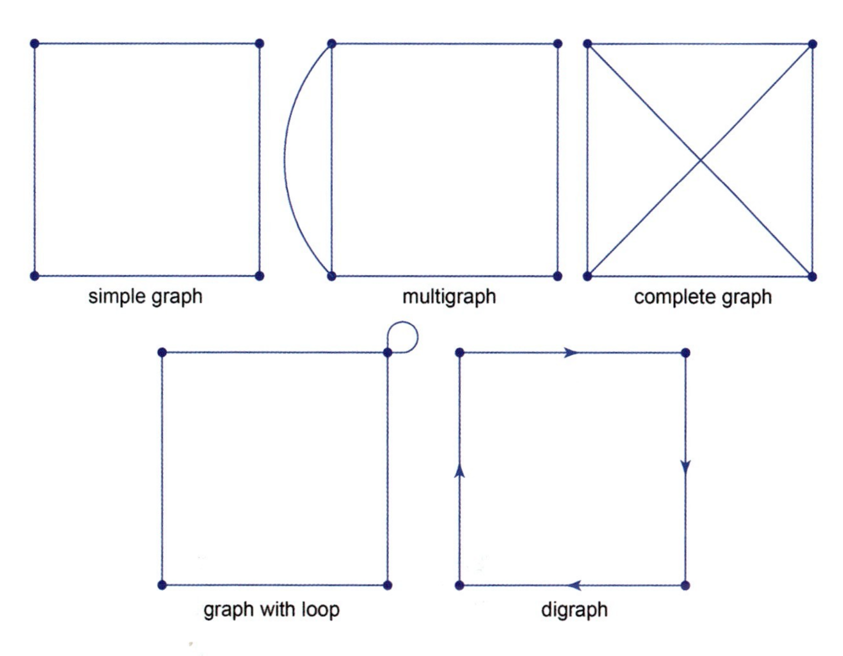

Basic types of graphs.

As used in graph theory, the term graph does not refer to data charts such as line graphs or bar graphs. Instead, it refers to a set of vertices (that is, points or nodes) and of edges (or lines) that connect the vertices. When any two vertices are joined by more than one edge, the graph is called a multigraph. A graph without loops and with at most one edge between any two vertices is called a simple graph. Unless stated otherwise, graph is assumed to refer to a simple graph. When each vertex is connected by an edge to every other vertex, the graph is called a complete graph. When appropriate, a direction may be assigned to each edge to produce what is known as a directed graph, or digraph.

An important number associated with each vertex is its degree; this is defined as the number of edges that enter or exit from it—thus, a loop contributes 2 to the degree of its vertex. For instance, the vertices of the simple graph shown in the diagram all have a degree of 2, whereas the vertices of the complete graph shown are all of degree 3. Knowing the number of vertices in a complete graph characterizes its essential nature. For this reason, complete graphs are commonly designated K_n, where n refers to the number of vertices, and all vertices of K_n have degree $n - 1$.

Another important concept in graph theory is the path, which is any route along the edges of a graph. A path may follow a single edge directly between two vertices, or it may follow multiple edges through multiple vertices. If there is a path linking any two vertices in a graph, that graph is said to be connected. A path that begins and ends at the same vertex without traversing any edge more than once is called a circuit, or a closed path. A circuit that follows each edge exactly once while visiting every vertex is known as an Eulerian circuit, and the graph is called an Eulerian graph. An Eulerian graph is connected and, in addition, all its vertices have even degree.

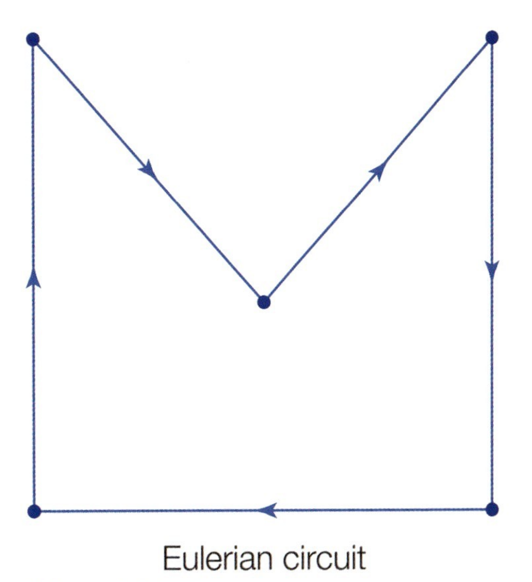

Eulerian circuit

A path that traverses each edge exactly once such that the path begins and ends at the same vertex is known as an Eulerian circuit.

In 1857 the Irish mathematician William Rowan Hamilton invented a puzzle ("The Icosian Game") that he later sold to a game manufacturer for £25. The puzzle involved finding a special type of path, later known as a Hamiltonian circuit, along the edges of a dodecahedron (a Platonic solid consisting of 12 pentagonal faces) that begins and ends at the same corner while passing through each corner exactly once. The knight's tour is another example of a recreational problem involving a Hamiltonian circuit.

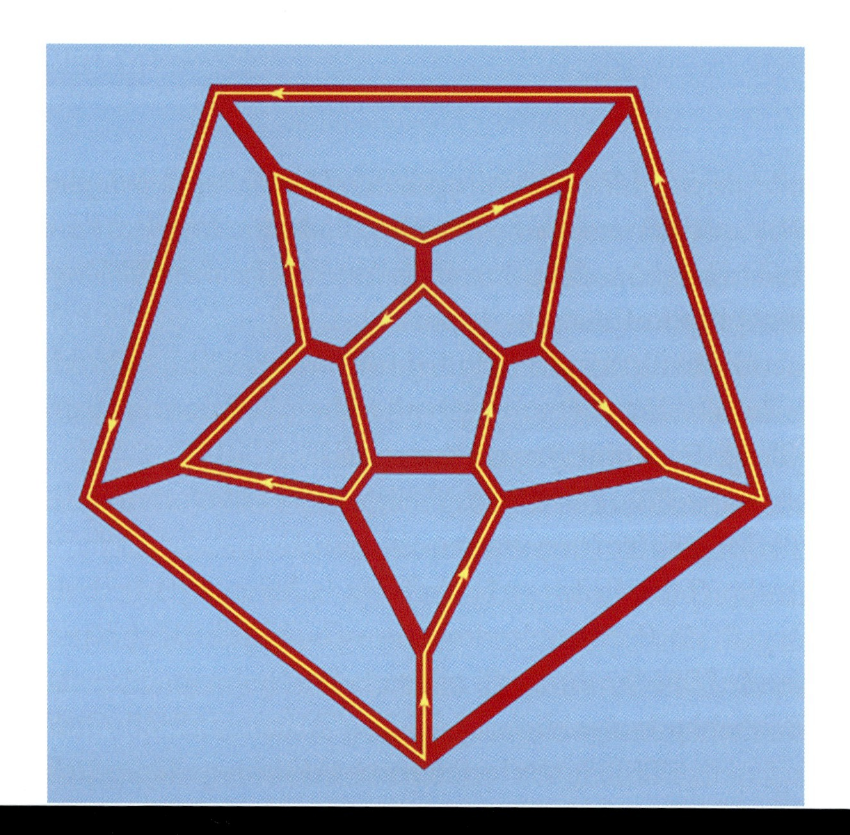

Hamiltonian circuit.

Hamiltonian graphs have been more challenging to characterize than Eulerian graphs, since the necessary and sufficient conditions for the existence of a Hamiltonian circuit in a connected graph are still unknown.

The histories of graph theory and topology are closely related, and the two areas share many common problems and techniques. Euler referred to his work on the Königsberg bridge problem as an example of *geometria situs*—the "geometry of position"—while the development of topological ideas during the second half of the

19th century became known as *analysis situs*—the "analysis of position." In 1750 Euler discovered the polyhedral formula $V - E + F = 2$ relating the number of vertices (V), edges (E), and faces (F) of a polyhedron (a solid, like the dodecahedron mentioned above, whose faces are polygons). The vertices and edges of a polyhedron form a graph on its surface, and this notion led to consideration of graphs on other surfaces such as a torus (the surface of a solid doughnut) and how they divide the surface into disklike faces. Euler's formula was soon generalized to surfaces as $V - E + F = 2 - 2g$, where g denotes the

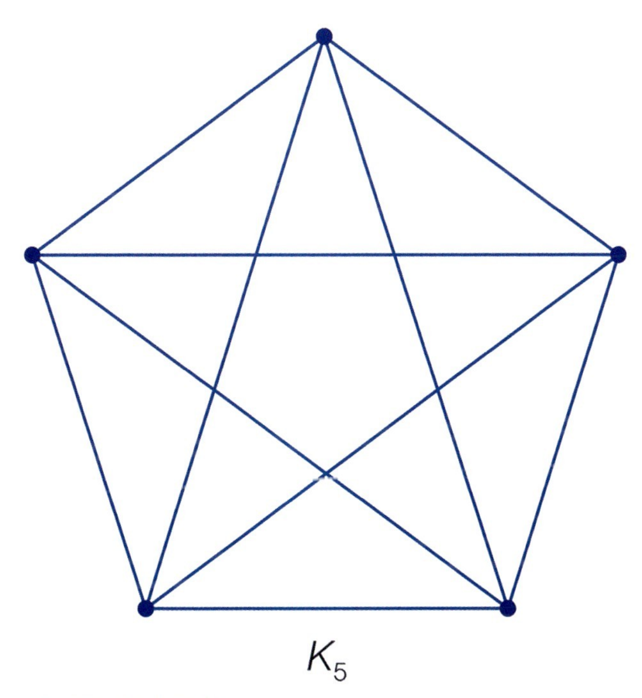

This is not a planar graph because there is no way to connect every vertex to every other vertex with edges in the plane such that no edges intersect.

genus, or number of "doughnut holes," of the surface. Having considered a surface divided into polygons by an embedded graph, mathematicians began to study ways of constructing surfaces, and later more general spaces, by pasting polygons together. This was the beginning of the field of combinatorial topology, which later, through the work of the French mathematician Henri Poincaré and others, grew into what is known as algebraic topology.

The connection between graph theory and topology led to a subfield called topological graph theory. An important problem in this area concerns planar graphs. These are graphs that can be drawn as dot-and-line diagrams on a plane (or equivalently, on a sphere) without any edges crossing except at the vertices where they meet. Complete graphs with four or fewer vertices are planar, but complete graphs with five vertices (K_5) or more are not. Nonplanar graphs cannot be drawn on a plane or on the surface of a sphere without edges intersecting each other between the vertices. The use of diagrams of dots and lines to represent graphs actually grew out of 19th-century chemistry, where lettered vertices denoted individual atoms and connecting lines denoted chemical bonds (with degree corresponding to valence), in which planarity had important chemical consequences. The first use, in this context, of the word graph is attributed to the 19th-century Englishman James Sylvester, one of several mathematicians interested in counting special types of diagrams representing molecules.

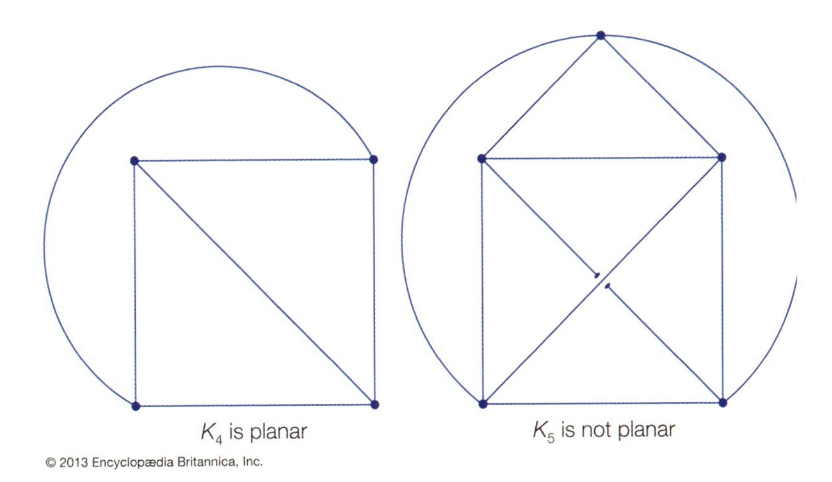

Only with fewer than five vertices in a two-dimensional plane can a collection of paths between each vertex be drawn such that no paths intersect.

Another class of graphs is the collection of the complete bipartite graphs $K_{m,n}$, which consist of the simple graphs that can be partitioned into two independent sets of m and n vertices, such that there are no edges between vertices within each set and every vertex in one set is connected by an edge to every vertex in the other set. Like K_5, the bipartite graph $K_{3,3}$ is not planar, disproving a claim made in 1913 by the English recreational problemist Henry Dudeney to a solution to the "gas-water-electricity problem." In 1930 the Polish mathematician Kazimierz Kuratowski proved that any nonplanar graph must contain a certain type of copy of K_5 or $K_{3,3}$. While K_5 and $K_{3,3}$ cannot be embedded

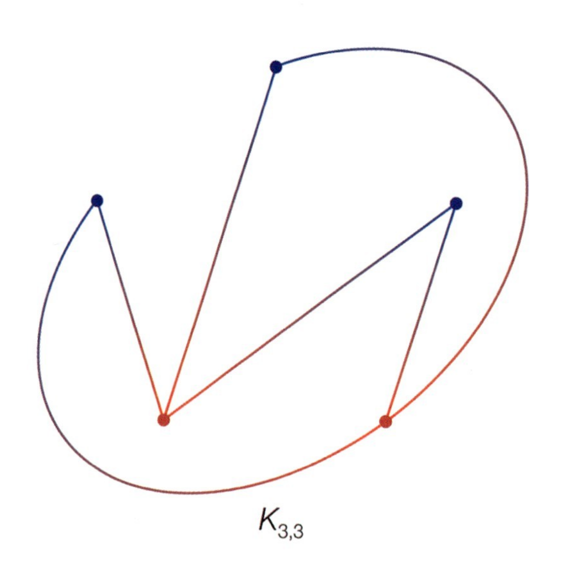

Bipartite map.

in the sphere, they can be embedded in a torus. The graph embedding problem concerns the determination of surfaces in which a graph can be embedded and thereby generalizes the planarity problem. It was not until the late 1960s that the embedding problem for the complete graphs K_n was solved for all n.

Another problem of topological graph theory is the map-colouring problem. This problem is an outgrowth of the well-known four-colour map problem, which asks whether the countries on every map can be coloured, using just four colours, in such a way that countries sharing an edge have different colours. Asked originally in the 1850s by Francis Guthrie, then a student at University College London, this problem has a rich history filled with incorrect attempts at its solution. In an equivalent graph theoretic form, one may translate this problem to ask whether the vertices of a planar graph can always be coloured, using just four colours, in such a way that vertices joined by an edge have different colours. The result was finally proved in 1976 utilizing computerized checking of nearly 2,000 special configurations. Interestingly, the corresponding colouring problem concerning the number of colours required to colour maps on surfaces of higher genus was

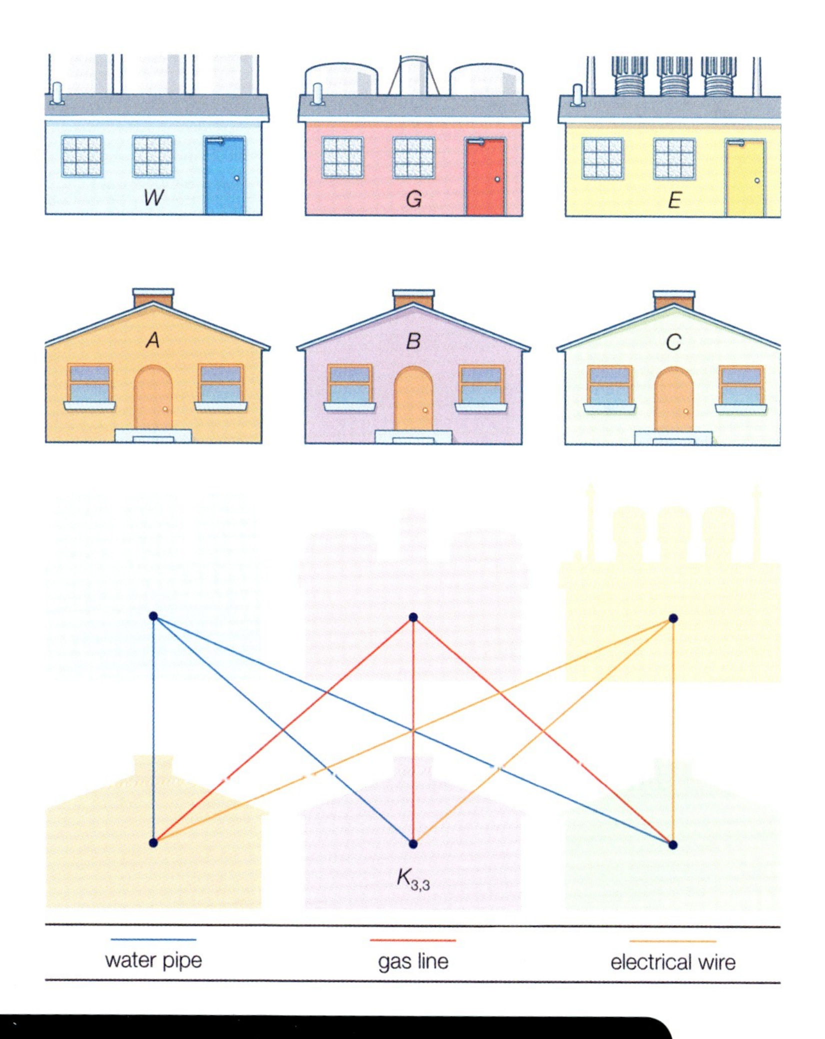

Dudeney puzzle.

completely solved a few years earlier—for example, maps on a torus may require as many as seven colours. This work confirmed that a formula of the English mathematician Percy Heawood from 1890 correctly gives these colouring numbers for all surfaces except the one-sided surface known as the Klein bottle, for which the correct colouring number had been determined in 1934.

Among the current interests in graph theory are problems concerning efficient algorithms for finding optimal paths (depending on different criteria) in graphs. Two well-known examples are the Chinese postman problem (the shortest path that visits each edge at least once), which was solved in the 1960s, and the traveling salesman problem (the shortest path that begins and ends at the same vertex and visits each edge exactly once), which continues to attract the attention of many researchers because of its applications in routing data, products, and people. Work on such problems is related to the field of linear programming, which was founded in the mid-20th century by the American mathematician George Dantzig.

Chapter 3

GEOMETRIC TERMS AND CONCEPTS

ALGEBRAIC SURFACE

In three-dimensional space, a surface the equation of which is $f(x, y, z) = 0$, with $f(x, y, z)$ a polynomial in x, y, z is called an algebraic surface. The order of the surface is the degree of the polynomial equation. If the surface is of the first order, it is a plane. If the surface is of order two, it is called a quadric surface. By rotating the surface, its equation can be put in the form

$$Ax^2 + By^2 + Cz^2 + Dx + Ey + Fz = G.$$

If A, B, C are all not zero, the equation can generally be simplified to the form

$$ax^2 + by^2 + cz^2 = 1.$$

This surface is called an ellipsoid if a, b, and c are positive. If one of the coefficients is negative, the surface is a hyperboloid of one sheet; if two of the coefficients are negative, the surface is a hyperboloid of two sheets. A hyperboloid of one sheet has a saddle point (a point on a curved surface shaped like a saddle at which the curvatures in two mutually perpendicular planes are of opposite signs, just like a saddle is curved up in one direction and down in another).

If A, B, C are possibly zero, then cylinders, cones, planes, and elliptic or hyperbolic paraboloids may be produced. Examples of the latter are $z = x^2 + y^2$ and $z = x^2 - y^2$, respectively. Through every point of a quadric pass two straight lines lying on the surface. A cubic surface is one of order three. It has the property that 27 lines lie on it, each one meeting 10 others. In general, a surface of order four or more contains no straight lines.

ANGLE TRISECTION: ARCHIMEDES' METHOD

Euclid's insistence (c. 300 BCE) on using only unmarked straightedge and compass for geometric constructions did not inhibit the imagination of his successors. Archimedes (c. 287–212/211 BCE) made use of *neusis* (the sliding and maneuvering of a measured length, or marked straightedge) to solve one of the great problems of ancient geometry: constructing an angle that is one-third the size of a given angle.

1. Given $\angle AOB$, draw the circle with centre at O through the points A and B. Thus, OA and OB are radii of the circle and $OA = OB$.
2. Extend the ray AO indefinitely.
3. Now take a straightedge marked with the length

of the circle's radius and maneuver it (this is the *neusis*) into position to draw a line segment from B through a point C on the circle to a point D on the ray AO such that CD is equal to the circle's radius; that is, $CD = OC = OB = OA$.

4. From the Bridge of Asses, $\angle CDO = \angle COD$ *and* $\angle OCB = \angle OBC$.
5. $\angle AOB = \angle ODC + \angle OBC$, because $\angle AOB$ is an angle external to ΔDOB and an external angle equals the sum of the opposite interior angles ($\angle AOB + \angle BOD = 180° = \angle BOD + \angle ODB + \angle DBO$).
6. $\angle OBC = \angle OCB$ (by step 4) $= \angle ODC + \angle COD$ (by step 5) $= 2\angle ODC$ (by step 4).
7. Substituting $2\angle ODC$ for $\angle OBC$ in step 5 and simplifying, $\angle AOB = 3\angle ODC$. Hence $\angle ODC$ is one-third the original angle, as required.

TRISECTING THE ANGLE: THE QUADRATRIX OF HIPPIAS

Hippias of Elis (fl. 5th century BCE) imagined a mechanical device to divide arbitrary angles into various proportions. His device depends on a curve, now known as the quadratrix of Hippias, that is produced by plotting the intersection of two moving line segments. Starting from a horizontal position, one segment is rotated at a constant rate through a right angle around one of its endpoints, while the second segment glides uniformly through a vertical distance equal to the first segment's length. Because both the angle rotation and the vertical displacement are produced by uniform motion, each moves through the same fraction of its entire journey in the same time. Hence, finding some proportion (say one-third) for a given angle (here $\angle COA$) is simple: find the equal proportion for vertical displacement of the point on the quadratrix at which the two segments intersect (C), locate the point (F) on the quadratrix at that height (one-third of the original height), and then draw the new angle ($\angle FOA$) through that point.

AXIOM

An axiom (also known as a postulate) is an indemonstrable first principle, rule, or maxim, that has found general acceptance or is thought worthy of common acceptance whether by virtue of a claim to intrinsic merit or on the basis of an appeal to self-evidence. An example would be: "Nothing can both be and not be at the same time and in the same respect."

In Euclid's *Elements* the first principles were listed in two categories, as postulates and as common notions. The former are principles of geometry and seem to have been thought of as required assumptions because their statement opened with "let there be demanded" (*ētesthō*). The common notions are evidently the same as what were termed "axioms" by Aristotle, who deemed axioms the first principles from which all demonstrative sciences must start; indeed Proclus, the last important Greek philosopher ("*On the First Book of Euclid*"), stated explicitly that the notion and axiom are synonymous. The principle distinguishing postulates from axioms, however, does not seem certain. Proclus debated various accounts of it, among them that postulates are peculiar to geometry whereas axioms are common either to all sciences that are concerned with quantity or to all sciences whatever.

In modern times, mathematicians have often used the words postulate and axiom as synonyms. Some recommend that the term axiom be reserved for the axioms of logic and postulate for those assumptions or first principles beyond the principles of logic by which a particular mathematical discipline is defined.

AXIOMATIC METHOD

The axiomatic method is a procedure by which an entire system (e.g., a science) is generated in accordance with specified rules by logical deduction from certain basic propositions (axioms or postulates), which in turn are constructed from a few terms taken as primitive. These terms and axioms may either be arbitrarily defined and constructed or else be conceived according to a model in which some intuitive warrant for their truth is felt to exist. The oldest examples of axiomatized systems are Aristotle's syllogistic and Euclid's geometry. Early in the 20th century the British philosophers Bertrand

Russell and Alfred North Whitehead attempted to formalize all of mathematics in an axiomatic manner. Scholars have even subjected the empirical sciences to this method, as J.H. Woodger has done in *The Axiomatic Method in Biology* (1937) and Clark Hull (for psychology) in *Principles of Behaviour* (1943).

BRACHISTOCHRONE

The brachistochrone is the planar curve on which a body subjected only to the force of gravity will slide (without friction) between two points in the least possible time. Finding the curve was a problem first posed by Galileo. In the late 17th century the Swiss mathematician Johann Bernoulli issued a challenge to solve this problem. He and his older brother Jakob, along with Gottfried Wilhelm Leibniz, Isaac Newton, and others, found the curve to be a cycloid.

BRIDGE OF ASSES

As previously noted, Euclid's fifth proposition in the first book of his *Elements* (that the base angles in an isosceles triangle are equal) may have been named the Bridge of Asses (Latin: *Pons Asinorum*) for medieval students who, clearly not destined to cross over into more abstract mathematics, had difficulty understanding the proof—or even the need for the proof. An alternative name for this famous theorem was Elefuga, which Roger

Bacon, writing around 1250, derived from Greek words indicating "escape from misery." Medieval schoolboys did not usually go beyond the Bridge of Asses, which thus marked their last obstruction before liberation from the *Elements*.

1. We are given that ΔABC is an isosceles triangle—that is, that $AB = AC$.
2. Extend sides AB and AC indefinitely away from A.
3. With a compass centred on A and open to a distance larger than AB, mark off AD on AB extended and AE on AC extended so that $AD = AE$.
4. $\angle DAC = \angle EAB$, because it is the same angle.
5. Therefore, $\Delta DAC \cong \Delta EAB$; that is, all the corresponding sides and angles of the two triangles are equal. By imagining one triangle to be superimposed on another, Euclid argued that the two are congruent if two sides and the included angle of one triangle are equal to the corresponding two sides and included angle of the other triangle (known as the side-angle-side theorem).
6. Therefore, $\angle ADC = \angle AEB$ and $DC = EB$, by step 5.
7. Now BD = CE because BD = $AD - AB$, $CE = AE - AC$, $AB = AC$, and $AD = AE$, all by construction.
8. $\Delta BDC \cong \Delta CEB$, by the side-angle-side theorem of step 5.
9. Therefore, $\angle DBC = \angle ECB$, by step 8.

10. Hence, $\angle ABC = \angle ACB$ because $\angle ABC = 180° - \angle DBC$ and $\angle ACB = 180° - \angle ECB$.

BROUWER'S FIXED POINT THEOREM

Brouwer's fixed point theorem of algebraic topology was stated and proved in 1912 by the Dutch mathematician L.E.J. Brouwer. Inspired by earlier work of the French mathematician Henri Poincaré, Brouwer investigated the behaviour of continuous functions mapping the ball of unit radius in n-dimensional Euclidean space into itself. In this context, a function is continuous if it maps close points to close points. Brouwer's fixed point theorem asserts that for any such function f there is at least one point x such that $f(x) = x$; in other words, such that the function f maps x to itself. Such a point is called a fixed point of the function.

When restricted to the one-dimensional case, Brouwer's theorem can be shown to be equivalent to the intermediate value theorem, which is a familiar result in calculus and states that if a continuous real-valued function f defined on the closed interval $[-1, 1]$ satisfies $f(-1) < 0$ and $f(1) > 0$, then $f(x) = 0$ for at least one number x between -1 and 1; less formally, an unbroken curve passes through every value between its endpoints. An n-dimensional version of the intermediate value theorem was shown to be equivalent to Brouwer's fixed point theorem in 1940.

There are many other fixed point theorems, including one for the sphere, which is the surface of a solid ball in three-dimensional space and to which Brouwer's theorem does not apply. The fixed point theorem for the sphere asserts that any continuous function mapping the sphere into itself either has a fixed point or maps some point to its antipodal point.

Fixed point theorems are examples of existence theorems, in the sense that they assert the existence of objects, such as solutions to functional equations, but not necessarily methods for finding such solutions. However, some of these theorems are coupled with algorithms that produce solutions, especially for problems in modern applied mathematics.

CATENARY

The catenary is a curve that describes the shape of a flexible hanging chain or cable—the name derives from the Latin *catenaria* ("chain"). Any freely hanging cable or string assumes this shape, also called a chainette, if the body is of uniform mass per unit of length and is acted upon solely by gravity.

Early in the 17th century, the German astronomer Johannes Kepler applied the ellipse to the description of planetary orbits, and the Italian scientist Galileo Galilei employed the parabola to describe projectile motion in the absence of air resistance. Inspired by the great success of conic sections in these settings, Galileo incorrectly believed that a hanging chain would take the shape of a parabola. It was later in the 17th century that the Dutch mathematician Christiaan Huygens showed that the chain curve cannot be given by an algebraic equation (one involving only arithmetic operations together with powers and roots); he also coined the term *catenary*. In addition to Huygens, the Swiss mathematician Jakob Bernoulli and the German mathematician Gottfried Leibniz contributed to the complete description of the equation of the catenary.

Precisely, the curve in the xy-plane of such a chain suspended from equal heights at its ends and dropping at $x = 0$ to its lowest height $y = a$ is given by the equation $y = (a/2)(e^{x/a} + e^{-x/a})$. It can also be expressed in terms of the hyperbolic cosine function as $y = a \cosh(x/a)$.

Although the catenary curve fails to be described by a parabola, it is of interest to note that it is related to a parabola: the curve traced in the plane by the focus of a parabola as it rolls along a straight line is a catenary. The surface of revolution generated when an upward-opening catenary is revolved around the horizontal axis is called a catenoid. The catenoid was discovered in 1744 by the Swiss mathematician Leonhard Euler and it is the only minimal surface, other than the plane, that can be obtained as a surface of revolution.

The catenary and the related hyperbolic functions play roles in other applications. An inverted hanging cable provides the shape for a stable self-standing arch, such as the Gateway Arch located in St. Louis, Missouri.

The hyperbolic functions also arise in the description of waveforms, temperature distributions, and the motion of falling bodies subject to air resistance proportional to the square of the speed of the body.

CEVA'S THEOREM

Ceva's theorem concerns the vertices and sides of a triangle. In particular, the theorem asserts that for a given triangle ABC and points L, M, and n that lie on the sides AB, BC, and CA, respectively, a necessary and sufficient condition for the three lines from vertex to point opposite (AM, BN, CL) to intersect at a common point (be concurrent) is that the following relation hold between the line segments formed on the triangle:

$$BM \cdot CN \cdot AL = MC \cdot NA \cdot LB.$$

Although the theorem is credited to the Italian mathematician Giovanni Ceva, who published its proof in *De Lineis Rectis* (1678; "*On Straight Lines*"), it was proved earlier by Yūsuf al-Muʾtamin, king (1081–85) of Saragossa. The theorem is quite similar to (technically, dual to) a geometric theorem proved by Menelaus of Alexandria in the 1st century CE.

CIRCLE

Geometrical curve, one of the conic sections, consisting of the set of all points the same distance (the radius) from a given point (the centre). A line connecting any two points on a circle is called a chord, and a chord passing through the centre is called a diameter. The distance around a circle (the circumference) equals the length of a diameter multiplied by π. The area of a circle is the square of the radius multiplied by π. An arc consists of any part of a circle encompassed by an angle with its vertex at the centre (central angle). Its length is in the same proportion to the circumference as the central angle is to a full revolution.

COMPACTNESS

Compactness is a property of some topological spaces (a generalization of Euclidean space) that has its main use in the study of functions defined on such spaces. An open covering of a space (or set) is a collection of open sets that covers the space; *i.e.*, each point of the space is in some member of the collection. A space is defined as being compact if from each such collection of open sets, a finite number of these sets can be chosen that also cover the space.

Formulation of this topological concept of compactness was motivated by the Heine-Borel theorem for Euclidean space, which states that compactness of a set is equivalent to the set's being closed and bounded.

In general topological spaces, there are no concepts of distance or boundedness; but there are some theorems concerning the property of being closed. In a Hausdorff space (*i.e.*, a topological space in which every two points can be enclosed in nonoverlapping open sets) every compact subset is closed, and in a compact space every closed subset is also compact. Compact sets also have the Bolzano-Weierstrass property, which means that for every infinite subset there is at least one point around which the other points of the set accumulate. In Euclidean space, the converse is also true; that is, a set having the Bolzano-Weierstrass property is compact.

Continuous functions on a compact set have the important properties of possessing maximum and minimum values and being approximated to any desired precision by properly chosen polynomial series, Fourier series, or various other classes of functions as described by the Stone-Weierstrass approximation theorem.

CONE

A cone is the surface traced by a moving straight line (the generatrix) that always passes through a fixed point (the vertex). The path, to be definite, is directed by some closed plane curve (the directrix), along which the line always glides. In a right circular cone, the directrix is a circle, and the cone is a surface of revolution. The axis of this cone is a line through the vertex and the centre of the circle, the line being perpendicular to the plane of the circle. In an oblique circular cone, the angle that the axis makes with the circle is other than 90°. The directrix of a cone need not be a circle; and if the cone is right, planes parallel to the plane of the directrix produce intersections with the cone that take the shape, but not the size, of the directrix. For such a plane, if the directrix is an ellipse, the intersection is an ellipse.

The generatrix of a cone is assumed to be infinite in length, extending in both directions from the vertex. The cone so generated, therefore, has two parts, called nappes or sheets, that extend infinitely. A finite cone has a finite, but not necessarily fixed, base, the surface enclosed by the directrix, and a finite, but not necessarily fixed, length of generatrix, called an element.

COORDINATE SYSTEMS

Arrangement of reference lines or curves used to identify the location of points in space. In two dimensions, the most common system is the Cartesian (after René Descartes) system. Points are designated by their distance along a horizontal (x) and vertical (y) axis from a reference point, the origin, designated (0, 0). Cartesian coordinates also can be used for three (or more) dimensions. A polar coordinate system locates a point by its direction relative to a reference direction and its distance from a given point, also the origin. Such a system is used in radar or sonar tracking and is the basis of bearing-and-range navigation systems. In three dimensions, it leads to cylindrical and spherical coordinates.

The coordinates are written (r, θ), in which r is the distance from the origin to any desired point and θ is the angle made by the vector r and the axis. A simple relationship exists between Cartesian coordinates given in terms of two reference axes (x, y) and the polar coordinates (r, θ), namely: $x = r \cos \theta$, and $y = r \sin \theta$.

Polar coordinates, like Cartesian coordinates, may also be used to locate points in three-dimensional space. The system used involves again the radius vector r, which gives distance from the origin, the angle θ, measured between r and the z-axis, and a second angle ϕ, measured between the x-axis and the projection of r in the x, y plane. This system is essentially identical to that of spherical coordinates; points on Earth, for example, are located in terms of latitude and longitude, which express angles measured with respect to the axis of the Earth's rotation and with respect to an arbitrary reference of longitude (the Greenwich meridian).

In a spherical coordinate system any point in three-dimensional space is specified by its angle with respect to a polar axis and angle of rotation with respect to a prime meridian on a sphere of a given radius. In spherical coordinates a point is specified by the triplet (r, θ, φ), where r is the point's distance from the origin (the radius), θ is the angle of rotation from the initial meridian plane, and φ is the angle from the polar axis (analogous to a ray from the origin through the North Pole).

CROSS RATIO

The cross ratio is of fundamental importance in characterizing projections. In a projection of one line onto another from a central point, the double ratio of lengths on the first line $(AC/AD)/(BC/BD)$ is equal to the corresponding ratio on the other line. Such a ratio is significant because projections distort most metric relationships (i.e., those involving the measured quantities of length and angle), while the study of projective geometry centres on finding those properties that remain invariant. Although the cross ratio was used extensively by early 19th-century projective geometers in formulating theorems, it was felt to be a somewhat unsatisfactory concept because its definition depended upon the Euclidean concept of length, a concept from which projective geometers wanted to free the subject altogether. In 1847 the German mathematician Karl G.C. von Staudt showed how to effect this separation by defining the cross ratio without reference to length. In 1873 the German mathematician Felix Klein showed how the basic concepts in Euclidean geometry of length and angle magnitude could be defined solely in terms of von Staudt's abstract cross ratio, bringing the two geometries together again, this time with projective geometry occupying the more basic position.

CURVE

Curve is an abstract term used to describe the path of a continuously moving point. Such a path is usually generated by an equation. The word can also apply to a straight line or to a series of line segments linked end to end. A closed curve is a path that repeats itself, and thus encloses one or more regions. Simple examples include circles, ellipses, and polygons. Open curves such as parabolas, hyperbolas, and spirals have infinite length.

CYCLOID

A cycloid is the curve generated by a point on the circumference of a circle that rolls along a straight line. If r is the radius of the circle and θ is the angular displacement of the circle, then the polar equations of the curve are $x = r(\theta - \sin \theta)$ and $y = r(1 - \cos \theta)$.

The points of the curve that touch the straight line are separated along the line by a distance equal to $2\pi r$, which is the circumference of the circle, indicating one complete revolution of the circle. The curve is periodic, which means that it repeats in an identical pattern for each cycle, or length of the line, that is equal to $2\pi r$.

One variant of the simple cycloid is the curtate cycloid, for which the curve falls below the line at the cusps, making retrograde loops in which the curve moves in the direction opposite to that of the rolling circle.

The prolate cycloid is similar to the simple cycloid except that the curve has no cusps and does not intersect the line. The prolate is formed by a point on a radius less than that of the rolling circle, such as a point on the spoke of a wheel.

For the case of a circle rolled along outside the circumference of another circle, an epicycloid is formed. For a circle rolled along inside the circumference of another circle, a hypocycloid is formed.

CYLINDER

A cyclinder is a surface of revolution that is traced by a straight line (the generatrix) that always moves parallel to itself or some fixed line or direction (the axis). The path, to be definite, is directed along a curve (the directrix), along which the line always glides. In a right circular cylinder, the directrix is a circle. The axis of this cylinder is a line through the centre of the circle, the line being perpendicular to the plane of the circle. In an oblique circular cylinder, the angle that the axis makes with the circle is other than 90°.

The directrix of a cylinder need not be a circle, and if the cylinder is right, planes parallel to the plane of the directrix that intersect the cylinder produce intersections that take the shape of the directrix. For such a plane, if the directrix is an ellipse, the intersection is an ellipse.

The generatrix of a cylinder is assumed to be infinite in length; the cylinder so generated, therefore, extends infinitely in both directions of its axis. A finite cylinder has a finite base, the surface enclosed by the directrix, and a finite length of generatrix, called an element.

DESARGUES'S THEOREM

Desargues's theorem, discovered by the French mathematician Girard Desargues in 1639, motivated the development, in the first quarter of the 19th century, of projective geometry by another French mathematician, Jean-Victor Poncelet. The theorem states that if two triangles ABC and A′B′C′, situated in three-dimensional space, are related to each other in such a way that they can be seen perspectively from one point (*i.e.*, the lines AA′, BB′, and CC′ all intersect in one point), then the points of intersection of corresponding sides all lie on one line, provided that no two corresponding sides are parallel. Should this last case occur, there will be only two points of intersection instead of three, and the theorem must be modified to include the result that these two points will lie on a line parallel to the two parallel sides of the triangles. Rather than modify the theorem to cover this special case, Poncelet instead modified Euclidean space itself by postulating points at infinity, which was the key for the development of projective geometry. In this new projective space (Euclidean space with added points at infinity), each straight line is given an added point at infinity, with parallel lines having a common point. After Poncelet discovered that Desargues's theorem could be more simply formulated in projective space, other theorems followed within this framework that could be stated more simply in terms of only intersections of lines and collinearity of points, with no need for reference to measures of distance, angle, congruence, or similarity.

DIMENSION

In common parlance, the measure of the size of an object, such as a box, usually given as length, width, and height, is called a dimension. In mathematics, the notion of dimension is an extension of the idea that a line is one-dimensional, a plane is two-dimensional, and space is three-dimensional. In mathematics and physics one also considers higher-dimensional spaces, such as four-dimensional space-time, where four numbers are needed to characterize a point: three to fix a point in space and one to fix the time. Infinite-dimensional spaces, first studied early in the 20th century, have played an increasingly important role both in mathematics and in parts of physics such as quantum field theory, where they represent the space of possible states of a quantum mechanical system.

In differential geometry one considers curves as one-dimensional, since a single number, or parameter, determines a point on a curve—for example, the distance, plus or minus, from a fixed point on the curve. A surface, such as the surface of the Earth, has two dimensions, since each point can be located by a pair of numbers—usually latitude and longitude. Higher-dimensional curved spaces were introduced by the German mathematician Bernhard Riemann in 1854 and have become both a major subject of study within mathematics and a basic component of modern physics, from Albert Einstein's theory of general relativity and the subsequent development of cosmological models of the universe to late-20th-century superstring theory.

In 1918 the German mathematician Felix Hausdorff introduced the notion of fractional dimension. This concept has proved extremely fruitful, especially in the hands of the Polish-French mathematician Benoit Mandelbrot, who coined the word *fractal* and showed how fractional dimensions could be useful in many parts of applied mathematics.

DUALITY

Duality is a principle in mathematics whereby one true statement can be obtained from another by merely interchanging two words. It is a property belonging to the branch of algebra known as lattice theory, which is involved with the concepts of order and structure common to different mathematical systems. A mathematical structure is called a lattice if it can be ordered in a specified way. Projective geometry, set theory, and symbolic logic are examples of systems with underlying lattice structures, and therefore also have principles of duality.

Projective geometry has a lattice structure that can be seen by ordering the points, lines, and planes by the inclusion relation. In the projective geometry of the plane, the words *point* and *line* can be interchanged, giving for example the dual statements: "Two points determine a line" and "Two lines determine a point." This last statement, sometimes false in Euclidean geometry, is always true in projective geometry because the axioms do not allow for parallel lines. Sometimes the language of a statement must be modified in order that the corresponding dual statement be clear. The dual of the statement "Two lines intersect in a point" is vague, while the dual of "Two lines determine a point" is clear. Even the statement "Two points intersect in a line," however, can be understood if a point is considered as a set (or "pencil") containing all the lines on which it lies, a concept itself dual to the idea of a line being considered as the set of all points that lie on it.

There is a corresponding duality in three-dimensional projective geometry between points and planes. Here, the line is its own dual, because it is determined by either two points or two planes.

In set theory, the relations "contained in" and "contains" can be interchanged, with the union becoming the intersection and vice-versa. In this case, the original structure remains unchanged, so it is called self-dual.

In symbolic logic there is a similar self-duality if "implied" and "is implied by" are interchanged, along with the logical connectives "and" and "or."

Duality, a pervasive property of algebraic structures, holds that two operations or concepts are interchangeable, all results holding in one formulation also holding in the other, the dual formulation.

The *Elements* Since The Middle Ages

With the European recovery and translation of Greek mathematical texts during the 12th century—the first Latin translation of Euclid's *Elements*, by Adelard of Bath, was made about 1120—and with the multiplication of universities beginning around 1200, the *Elements* was installed as the ultimate textbook in Europe. Academic demand made it attractive to printers, and soon vernacular versions were introduced throughout Europe: the first English translation was made by Sir Henry Billingsley in 1570. However, despite availability of the *Elements* and repeated endorsement of the usefulness of geometry in exercising the reason and improving the arts and sciences, no more of it was taught in many secondary and higher schools in early modern Europe than in the Dark Ages.

In 1662 the famous diarist Samuel Pepys, then a senior official of the British Admiralty, had to hire a tutor to teach him the multiplication table. He had no arithmetic, let alone geometry, although he had received a bachelor's and a master's degree from Magdalene College, University of Cambridge. Beginning in the 18th century, however, owing to interest in Isaac Newton's physics and the need for more accurate navigation, mathematics improved in England. The *Elements* became the kernel of the most prestigious course of study at Cambridge, and Euclidean proofs were formalized so that each assertion and its justification came on a separate line. As a wider proportion of the populace obtained a secondary education in the later 19th century, geometry courses departed from slavish dependence on Euclid, despite strong opposition from traditionalists like Lewis Carroll, the Oxford don who wrote *Alice in Wonderland*.

This freer approach had long been followed on the Continent. The Jesuits, the schoolmasters of Europe during

(Continued on the next page)

(Continued from the previous page)

the 17th and most of the 18th century, took liberties in drumming geometry into non-mathematical heads. Jesuit professors of mathematics rearranged the *Elements*, added a little algebra, and dropped propositions and proofs deemed irrelevant or useless. This *lèse-majesté* was carried farthest in France, which, perhaps in consequence, produced the largest number of good geometers in Europe during the late 18th century. One of them, Adrien-Marie Legendre, produced a version of the *Elements* that had an immense influence. He used trigonometry, eliminated Euclid's wearisome treatment of incommensurables, omitted proofs of the obvious, and added practical examples. His approach was incorporated into the curriculum of the secondary schools (lycées) devised during the French Revolution. Translations and adaptations of French geometry textbooks invaded American high schools and colleges. A leading U.S. textbook in 1890 was the 42nd edition of Legendre's *Elements Americanized*.

The 20th century saw an accelerating move away from Euclid's form of teaching geometry by rigorously and systematically building up the subject. Proportionally more individuals studying geometry, accompanied by a general decline in teaching standards, recommended simplification. More algebra, elementary trigonometry, analytical geometry, and problems for pocket calculators obscured what remained of Euclid's method. As the (British) Mathematical Association declared in 1923, apropos the replacement of geometrical argument by trigonometry, "human nature takes refuge only too readily in a formula." The pocket calculator and the personal computer may hold the key to the way back. Outlaw the calculator and employ the computer, which invites attention to images and makes possible the easy manipulation of diagrams, for advancing understanding of geometrical relationships.

ELLIPSE

The intersection of a right circular cone and a plane that is not parallel to the base, the axis, or an element of the cone produces what is known as an ellipse. It may be defined as the path of a point moving in a plane so that the ratio of its distances from a fixed point (the focus) and a fixed straight line (the directrix) is a constant less than one.

Any such path has this same property with respect to a second fixed point and a second fixed line, and ellipses often are regarded as having two foci and two directrixes. The ratio of distances, called the eccentricity, is the discriminant (of a general equation that represents all the conic sections). Another definition of an ellipse is that it is the locus of points for which the sum of their distances from two fixed points (the foci) is constant. The smaller the distance between the foci, the smaller is the eccentricity and the more closely the ellipse resembles a circle.

A straight line drawn through the foci and extended to the curve in either direction is the major diameter (or major axis) of the ellipse. Perpendicular to the major axis through the centre, at the point on the major axis equidistant from the foci, is the minor axis. A line drawn through either focus parallel to the minor axis is a latus rectum (literally, "straight side").

The ellipse is symmetrical about both its axes. The curve when rotated about either axis forms the surface called the ellipsoid of revolution, or a spheroid.

The path of a heavenly body moving around another in a closed orbit in accordance with Newton's gravitational law is an ellipse. In the solar system one focus of such a path about the Sun is the Sun itself.

For an ellipse the centre of which is at the origin and the axes of which are coincident with the x and y axes, the equation is $x^2/a^2 + y^2/b^2 = 1$. The length of the major diameter is $2a$; the length of the minor diameter is $2b$. If c is taken as the distance from the origin to the focus, then $c^2 = a^2 - b^2$, and the foci of the curve may be located when the major and minor diameters are known. The problem of finding an exact expression for the perimeter of an ellipse led to the development of elliptic functions, an important topic in mathematics and physics.

ELLIPSOID

An ellipsoid is a closed surface of which all plane cross sections are either ellipses or circles. An ellipsoid is symmetrical about three mutually perpendicular axes that intersect at the centre.

If a, b, and c are the principal semiaxes, the general equation of such an ellipsoid is $x^2/a^2 + y^2/b^2 + z^2/c^2 = 1$. A special case arises when $a = b = c$: then the surface is a sphere, and the intersection with any plane passing through it is a circle. If two axes are equal, say $a = b$, and different from the third, c, then the ellipsoid is an ellipsoid of revolution, or spheroid, the figure formed by revolving an ellipse about one of its axes. If a and b are greater than c, the spheroid is oblate; if less, the surface is a prolate spheroid.

An oblate spheroid is formed by revolving an ellipse about its minor axis; a prolate, about its major axis. In either case, intersections of the surface by planes parallel to the axis of revolution are ellipses, while intersections by planes perpendicular to that axis are circles.

Isaac Newton predicted that because of the Earth's rotation, its shape should be an ellipsoid rather than spherical, and careful measurements confirmed his prediction. As more accurate measurements became possible, further deviations from the elliptical shape were discovered.

Often an ellipsoid of revolution (called the reference ellipsoid) is used to represent the Earth in geodetic calculations, because such calculations are simpler than those with more complicated mathematical models. For this ellipsoid, the difference between the equatorial radius and the polar radius (the semimajor and semiminor axes, respectively) is about 21 km (13 miles), and the flattening is about 1 part in 300.

Envelope

An envelope is a curve that is tangential to each one of a family of curves in a plane or, in three dimensions, a surface that is tangent to each one of a family of surfaces. For example, two parallel lines are the envelope of the family of circles of the same radius having centres on a straight line. An example of the envelope of a family of surfaces in space is the circular cone $x^2 - y^2 = z^2$ as the envelope of the family of paraboloids $x^2 + y^2 = 4a(z - a)$.

EUCLID'S WINDMILL

The Pythagorean theorem states that the sum of the squares on the legs of a right triangle is equal to the square on the hypotenuse (the side opposite the right angle)—in familiar algebraic notation, $a^2 + b^2 = c^2$. The Babylonians and Egyptians had found some integer triples (a, b, c) satisfying the relationship. Pythagoras (c. 580–c. 500 BCE) or one of his followers may have been the first to prove the theorem that bears his name. Euclid (c. 300 BCE) offered a clever demonstration of the Pythagorean theorem in his *Elements*, known as the Windmill proof from the figure's shape.

1. Draw squares on the sides of the right ΔABC.
2. BCH and ACK are straight lines because $\angle ACB = 90°$.
3. $\angle EAB = \angle CAI = 90°$, by construction.
4. $\angle BAI = \angle BAC + \angle CAI = \angle BAC + \angle EAB = \angle EAC$, by 3.
5. $AC = AI$ and $AB = AE$, by construction.

6. Therefore, $\Delta BAI \cong \Delta EAC$, by the side-angle-side theorem.
7. Draw CF parallel to BD.
8. Rectangle $AGFE = 2\Delta ACE$. This remarkable result derives from two preliminary theorems: (a) the areas of all triangles on the same base, whose third vertex lies anywhere on an indefinitely extended line parallel to the base, are equal; and (b) the area of a triangle is half that of any parallelogram (including any rectangle) with the same base and height.
9. Square $AIHC = 2\Delta BAI$, by the same parallelogram theorem as in step 8.
10. Therefore, rectangle $AGFE$ = square $AIHC$, by steps 6, 8, and 9.
11. $\angle DBC = \angle ABJ$, as in steps 3 and 4.
12. $BC = BJ$ and $BD = AB$, by construction as in step 5.
13. $\Delta CBD \cong \Delta JBA$, as in step 6.
14. Rectangle $BDFG = 2\Delta CBD$, as in step 8.
15. Square $CKJB = 2\Delta JBA$, as in step 9.
16. Therefore, rectangle $BDFG$ = square $CKJB$, as in step 10.
17. Square $ABDE$ = rectangle $AGFE$ + rectangle $BDFG$, by construction.
18. Therefore, square $ABDE$ = square $AIHC$ + square $CKJB$, by steps 10 and 16.

The first book of Euclid's *Elements* begins with the definition of a point and ends with the Pythagorean theorem and its converse (if the sum of the squares on two sides of a triangle equals the square on the third side, it must be a right triangle). This journey from particular definition to abstract and universal mathematical statement has been taken as emblematic of the development of civilized life. A striking example of the identification of Euclid's reasoning with the highest expression of thought was the proposal made in 1821 by a German physicist and astronomer to open a conversation with the inhabitants of Mars by showing them our claims to intellectual maturity. All we needed to do to attract their interest and approbation, it was claimed, was to plow and plant large fields in the shape of the windmill diagram or, as others proposed, to dig canals suggestive of the Pythagorean theorem in Siberia or the Sahara, fill them with oil, set them on fire, and await a response. The experiment has not been tried, leaving undecided whether the inhabitants of Mars have no telescope, no geometry, or no existence.

Euclidean Space

In geometry, Euclidean space is a two- or three-dimensional space in which the axioms and postulates of Euclidean geometry apply. It is also a space in any finite number of dimensions, in which points are designated by coordinates (one for each dimension) and the distance between two points is given by a distance formula. The only conception of physical space for over 2,000 years, it remains the most compelling and useful way of modeling the world as it is experienced. Though non-Euclidean spaces, such as those that emerge from elliptic geometry and hyperbolic geometry, have led scientists to a better understanding of the universe and of mathematics itself, Euclidean space remains the point of departure for their study.

Method of Exhaustion

The method of exhaustion is a technique invented by the classical Greeks to prove propositions regarding the areas and volumes of geometric figures. Although it was a forerunner of the integral calculus, the method of exhaustion used neither limits nor arguments about infinitesimal quantities. It was instead a strictly logical procedure, based upon the axiom that a given quantity can be made smaller than another given quantity by successively halving it (a finite number of times). From this axiom it can be shown, for example, that the area of a circle is proportional to the square of its radius. The term method of exhaustion was coined in Europe after the Renaissance and applied to the rigorous Greek procedures as well as to contemporary "proofs" of area formulas by "exhausting" the area of figures with successive polygonal approximations.

Fractal

A fractal is any of a class of complex geometric shapes that commonly have "fractional dimension," a concept first introduced by the mathematician Felix Hausdorff in 1918. Fractals are distinct from the simple figures of classical, or Euclidean, geometry—the square, the circle, the sphere, and so forth. They are capable of describing many irregularly shaped objects or spatially nonuniform phenomena in nature such as coastlines and mountain ranges. The term *fractal*, derived from the Latin word *fractus* ("fragmented," or "broken"), was coined by the Polish-born mathematician Benoit B. Mandelbrot. See the animation of the Mandelbrot fractal set.

Although the key concepts associated with fractals had been studied for years by mathematicians, and many examples, such as the Koch or "snowflake" curve were long known, Mandelbrot was the first to point out that fractals could be an ideal tool in applied mathematics for modeling a variety of phenomena from physical objects to the behavior of the stock market. Since its introduction in 1975, the concept of the fractal has given rise to a new system of geometry that has had a significant impact on such diverse fields as physical chemistry, physiology, and fluid mechanics.

Many fractals possess the property of self-similarity, at least approximately, if not exactly. A self-similar object is one whose component parts resemble the whole. This reiteration of details or patterns occurs at progressively smaller scales and can, in the case of purely abstract entities, continue indefinitely, so that each part of each part, when magnified, will look basically like a fixed part of the whole object. In effect, a self-similar object remains invariant under changes of scale—i.e., it has scaling symmetry. This fractal phenomenon can often be detected in such objects as snowflakes and tree barks. All natural fractals of this kind, as well as some mathematical self-similar ones, are stochastic, or random; they thus scale in a statistical sense.

Another key characteristic of a fractal is a mathematical parameter called its fractal dimension. Unlike Euclidean dimension, fractal dimension is generally expressed by a noninteger—that is to say, by a fraction rather than by a whole number. Fractal dimension can be illustrated by considering a specific example: the snowflake curve defined by Helge von Koch in 1904. It is a purely mathematical figure with a six-fold symmetry, like a natural snowflake. It is self-similar in that it consists of three identical parts, each of which in turn is made of four parts that are exact scaled-down versions of the whole. It follows that each of the four parts itself consists of four parts that are-scaled down versions of the whole. There would be nothing surprising if the scaling factor were also four, since that would be true of a line segment or a circular arc. However, for the snowflake curve, the scaling factor at each stage is three. The fractal dimension, *D*, denotes the power to which 3 must be raised to produce 4—i.e., $3^D = 4$. The dimension of the snowflake curve is thus $D = {}^{\log 4}/_{\log 3}$, or roughly 1.26. Fractal dimension is a key property and an indicator of the complexity of a given figure.

Fractal geometry with its concepts of self-similarity and noninteger dimensionality has been applied increasingly in statistical mechanics, notably when dealing with physical systems consisting of seemingly random features. For example, fractal simulations have been used to plot the distribution of galaxy clusters throughout the universe and to study problems related to fluid turbulence. Fractal geometry also has contributed to computer graphics. Fractal algorithms have made it possible to generate lifelike images of complicated, highly irregular natural objects, such as the rugged terrains of mountains and the intricate branch systems of trees.

GOLDEN RATIO

The golden ratio, which has also been known as the golden section, golden mean, or divine proportion, is the irrational number $(1 + \sqrt{5})/2$, often denoted by the Greek letters τ or ϕ, and approximately equal to 1.618. The origin of this number and its name may be traced back to about 500 BCE and the investigation in Pythagorean geometry of the regular pentagon, in which the five diagonals form a five-pointed star. On each such diagonal lie two points of intersection with other diagonals, and either of those points divides the whole diagonal into two segments of unequal lengths so that the ratio of the whole diagonal to the larger segment equals the ratio of the larger segment to the smaller one. In terms of present day algebra, letting the length of the shorter segment be one unit and the length of the larger segment be x units gives rise to the equation $(x + 1)/x = x/1$; this may be rearranged to form the quadratic equation $x^2 - x - 1 = 0$, for which the positive solution is $x = (1 + \sqrt{5})/2$, the golden ratio.

The ancient Greeks recognized this "dividing" or "sectioning" property and described it generally as "the division of a line into extreme and mean ratio," a phrase that was ultimately shortened to simply "the section." It was more than 2,000 years later that both "ratio" and "section" were designated as "golden" in references by the German astronomer Johannes Kepler and others. The Greeks also had observed that the golden ratio provided the most aesthetically pleasing proportion of sides of a rectangle, a notion that was enhanced during the Renaissance by, for example, work of the Italian polymath Leonardo da Vinci and the publication of *De divina proportione* (1509; *Divine Proportion*) by the Italian mathematician Luca Pacioli, and illustrated by Leonardo.

The golden ratio occurs in many mathematical contexts. It is geometrically constructible by straightedge and compass, and it occurs in the investigation of the Archimedean and Platonic solids. It is the limit of the ratios of consecutive terms of the Fibonacci number sequence 1, 1, 2, 3, 5, 8, 13, …, in which each term beyond the second is the sum of the previous two, and it is also the value of the most basic of continued fractions, namely 1 + 1/(1 + 1/(1 + 1/(1 + ….

In modern mathematics, the golden ratio occurs in the description of fractals, figures that exhibit self-similarity and play an important role in the study of chaos and dynamical systems.

GRAPH

A graph is a pictorial representation of statistical data or of a functional relationship between variables. Graphs have the advantage of showing general tendencies in the quantitative behaviour of data, and therefore serve a predictive function. As mere approximations, however, they can be inaccurate and sometimes misleading.

Most graphs employ two axes, in which the horizontal axis represents a group of independent variables, and the vertical axis represents a group of dependent variables. The most common graph is a broken-line graph, where the independent variable is usually a factor of time. Data points are plotted on such a grid and then connected with line segments to give an approximate curve of, for example, seasonal fluctuations in sales trends. Data points need not be connected in a broken line, however. Instead they may be simply clustered around a median line or curve, as is often the case in experimental physics or chemistry.

If the independent variable is not expressly temporal, a bar graph may be used to show discrete numerical quantities in relation to each other. To illustrate the relative populations of various nations, for example, a series of parallel columns, or bars, may be used. The length of each bar would be proportional to the size of the population of the respective country it represents. Thus, a demographer could see at a glance that China's population is about 30 percent larger than its closest rival, India.

This same information may be expressed in a part-to-whole relationship by using a circular graph, in which a circle is divided into sections, and where the size, or angle, of each sector is directly proportional to the percentage of the whole it represents. Such a graph would show the same relative population sizes as the bar graph, but it would also illustrate that approximately one-fourth of the world's population resides in China. This type of graph, also known as a pie chart, is most commonly used to show the breakdown of items in a budget.

In analytic geometry, graphs are used to map out functions of two variables on a Cartesian coordinate system, which is composed of a horizontal x-axis, or abscissa, and a vertical y-axis, or ordinate. Each axis is a real number line, and their intersection at the zero point of each is called the origin. A graph in this sense is the locus of all points (x,y) that satisfy a particular function.

The easiest functions to graph are linear, or first-degree, equations, the simplest of which is $y = x$. The graph of this equation is a straight line that traverses the lower left and upper right quadrants of the graph, passing through the origin at a 45-degree angle.

Such regularly shaped curves as parabolas, hyperbolas, circles, and ellipses are graphs of second-degree equations. These and other nonlinear functions are sometimes graphed on a logarithmic grid, where a point on an axis is not the variable itself but the logarithm of that variable. Thus, a parabola with Cartesian coordinates may become a straight line with logarithmic coordinates.

In certain cases, polar coordinates provide a more appropriate graphic system, whereby a series of concentric circles with straight lines through their common centre, or origin, serves to locate points on a circular plane. Both Cartesian and polar coordinates may be expanded to represent three dimensions by introducing a third variable into the respective algebraic or trigonometric functions. The inclusion of three axes results in an isometric graph for solid bodies in the former case and a graph with spherical coordinates for curved surfaces in the latter.

HARMONIC CONSTRUCTION

Harmonic construction is a process of determining a pair of points C and D that divide a line segment AB harmonically, that is, internally and externally in the same ratio, the internal ratio CA/CB being equal to the negative of the external ratio DA/DB on the extended line. The theorem of harmonicity states that if the external point of division of a line segment is given, then the internal point can be constructed by a purely projective technique; that is, by using only intersections of straight lines. To accomplish this, an arbitrary triangle is drawn on the base AB, followed by an arbitrary line from the external point D cutting this triangle in two. The corners of the quadrilateral formed thus joined and the point determined by the intersection of these diagonals together with the point at the vertex of the triangle determine a line that cuts AB in the proper ratio.

This construction is of interest in projective geometry because the location of the fourth point is independent of the choice of the first three lines in the construction, and the harmonic relationship of the four points is preserved if the line is projected onto another line.

HAUSDORFF SPACE

A Hausdorff space is a type of topological space named for the German mathematician Felix Hausdorff. A topological space is a generalization of the notion of an object in three-dimensional space. It consists of an abstract set of points along with a specified collection of subsets, called open sets, that satisfy three axioms: (1) the set itself and the empty set are open sets, (2) the intersection of a finite number of open sets is open, and (3) the union of any collection of open sets is an open set. A Hausdorff space is a topological space with a separation property: any two distinct points can be separated by disjoint open sets—that is, whenever p and q are distinct points of a set X, there exist disjoint open sets Up and Uq such that $U p$ contains p and Uq contains q.

The real number line becomes a topological space when a set U of real numbers is declared to be open if and only if for each point p of U there is an open interval centred at p and of positive (possibly very small) radius completely contained in U. Thus, the real line also becomes a Hausdorff space since two distinct points p and q, separated a positive distance r, lie in the disjoint open intervals of radius $r/2$ centred at p and q, respectively. A similar argument confirms that any metric space, in which open sets are induced by a distance function, is a Hausdorff space. However, there are many examples of non-Hausdorff topological spaces, the simplest of which is the trivial topological space consisting of a set x with at least two points and just x and the empty set as the open sets. Hausdorff spaces satisfy many properties not satisfied generally by topological spaces. For example, if two continuous functions f and g map the real line into a Hausdorff space and $f(x) = g(x)$ for each rational number x, then $f(x) = g(x)$ for each real number x.

Hausdorff included the separation property in his axiomatic description of general spaces in *Grundzüge der Mengenlehre* (1914; "*Elements* of Set Theory"). Although later it was not accepted as a basic axiom for topological spaces, the Hausdorff property is often assumed in certain areas of topological research. It is one of a long list of properties that have become known as "separation axioms" for topological spaces.

Hilbert Space

A Hilbert space is an example of an infinite-dimensional space that had a major impact in analysis and topology. The German mathematician David Hilbert first described this space in his work on integral equations and Fourier series, which occupied his attention during the period 1902–12.

The points of Hilbert space are infinite sequences $(x_1, x_2, x_3, \ldots)$ of real numbers that are square summable, that is, for which the infinite series $x_1^2 + x_2^2 + x_3^2 + \ldots$ converges to some finite number. In direct analogy with n-dimensional Euclidean space, Hilbert space is a vector space that has a natural inner product, or dot product, providing a distance function. Under this distance function it becomes a complete metric space and, thus, is an example of what mathematicians call a complete inner product space.

Soon after Hilbert's investigation, the Austrian-German mathematician Ernst Fischer and the Hungarian mathematician Frigyes Riesz proved that square integrable functions (functions such that integration of the square of their absolute value is finite) could also be considered as "points" in a complete inner product space that is equivalent to Hilbert space. In this context, Hilbert space played a role in the development of quantum mechanics, and it has continued to be an important mathematical tool in applied mathematics and mathematical physics.

In analysis, the discovery of Hilbert space ushered in functional analysis, a new field in which mathematicians study the properties of quite general linear spaces. Among these spaces are the complete inner product spaces, which now are called Hilbert spaces, a designation first used in 1929 by the Hungarian-American mathematician John von Neumann to describe these spaces in an abstract axiomatic way. Hilbert space has also provided a source for rich ideas in topology. As a metric space, Hilbert space can be considered an infinite-dimensional linear topological space, and important questions related to its topological properties were raised in the first half of the 20th century. Motivated initially by such properties of Hilbert spaces, researchers established a new subfield of topology called infinite dimensional topology in the 1960s and '70s.

HIPPOCRATES' QUADRATURE OF THE LUNE

Hippocrates of Chios (fl. c. 460 BCE) demonstrated that the moon-shaped areas between circular arcs, known as lunes, could be expressed exactly as a rectilinear area, or quadrature. In the following simple case, two lunes developed around the sides of a right triangle have a combined area equal to that of the triangle.

1. Starting with the right ΔABC, draw a circle whose diameter coincides with AB (side c), the hypotenuse. Because any right triangle drawn with a cir-

cle's diameter for its hypotenuse must be inscribed within the circle, C must be on the circle.

2. Draw semicircles with diameters AC (side b) and BC (side a) as in the figure.
3. Label the resulting lunes L_1 and L_2 and the resulting segments S_1 and S_2, as indicated in the figure.
4. Now the sum of the lunes (L_1 and L_2) must equal the sum of the semicircles ($L_1 + S_1$ and $L_2 + S_2$) containing them minus the two segments (S_1 and S_2). Thus, $L_1 + L_2 = \pi/2(b/2)^2 - S_1 + \pi/2(a/2)^2 - S_2$ (since the area of a circle is π times the square of the radius).
5. The sum of the segments (S_1 and S_2) equals the area of the semicircle based on AB minus the area of the triangle. Thus, $S_1 + S_2 = \pi/2(c/2)^2 - \Delta ABC$.
6. Substituting the expression in step 5 into step 4 and factoring out common terms, $L_1 + L_2 = \pi/8(a^2 + b^2 - c^2) + \Delta ABC$.
7. Since $\angle ACB = 90°$, $a^2 + b^2 - c^2 = 0$, by the Pythagorean theorem. Thus, $L_1 + L_2 = \Delta ABC$.

Hippocrates managed to square several sorts of lunes, some on arcs greater and less than semicircles, and he intimated, though he may not have believed, that his method could square an entire circle. At the end of the classical age, Boethius (c. 470–524), whose Latin translations of snippets of Euclid would keep the light of geometry flickering for half a millennium, mentioned that someone had accomplished the squaring of the circle.

Whether the unknown genius used lunes or some other method is not known, since for lack of space Boethius did not give the demonstration. He thus transmitted the challenge of the quadrature of the circle together with fragments of geometry apparently useful in performing it. Europeans kept at the hapless task well into the Enlightenment. Finally, in 1775, the Paris Academy of Sciences, fed up with the task of spotting the fallacies in the many solutions submitted to it, refused to have anything further to do with circle squarers.

HYPERBOLA

A hyperbola is a type of conic section, a two-branched open curve produced by the intersection of a circular cone and a plane that cuts both nappes of the cone. As a plane curve it may be defined as the path (locus) of a point moving so that the ratio of the distance from a fixed point (the focus) to the distance from a fixed line (the directrix) is a constant greater than one. The hyperbola, however, because of its symmetry, has two foci. Another definition is that of a point moving so that the difference of its distances from two fixed points, or foci, is a constant. A degenerate hyperbola (two intersecting lines) is formed by the intersection of a circular cone and a plane that cuts both nappes of the cone through the apex.

A line drawn through the foci and prolonged beyond is the transverse axis of the hyperbola; perpendicular to that axis, and intersecting it at the geometric centre of the hyperbola, a point midway between the two foci, lies the conjugate axis. The hyperbola is symmetrical with respect to both axes.

Two straight lines, the asymptotes of the curve, pass through the geometric centre. The hyperbola does not intersect the asymptotes, but its distance from them becomes arbitrarily small at great distances from the centre. The hyperbola when revolved about either axis forms a hyperboloid.

For a hyperbola that has its centre at the origin of a Cartesian coordinate system and has its transverse axis lying on the x axis, the coordinates of its points satisfy the equation $x^2/a^2 - y^2/b^2 = 1$, in which a and b are constants.

HYPERBOLOID

A hyperboloid is the open surface generated by revolving a hyperbola about either of its axes. If the tranverse axis of the surface lies along the x axis and its centre lies at the origin and if a, b, and c are the principal semi-axes, then the general equation of the surface is expressed as $x^2/a^2 \pm y^2/b^2 - z^2/c^2 = 1$.

Revolution of the hyperbola about its conjugate axis generates a surface of one sheet, an hourglass-like shape, for which the second term of the above equation is positive. The intersections of the surface with planes parallel to the xz and yz planes are hyperbolas. Intersections with planes parallel to the xy plane are circles or ellipses.

Revolution of the hyperbola about its transverse axis generates a surface of two sheets, two separate surfaces, for which the second term of the general equation is negative. Intersections of the surface(s) with planes parallel to the *xy* and *xz* planes produce hyperbolas. Cutting planes parallel to the *yz* plane and at a distance greater than the absolute value of a, $|a|$, from the origin produce circles or ellipses of intersection, respectively, as a equals b or a is not equal to b.

INCOMMENSURABLES

The geometers immediately following Pythagoras (c. 580–c. 500 BCE) shared the unsound intuition that any two lengths are "commensurable" (that is, measurable) by integermultiples of some common unit. To put it another way, they believed that the whole (or counting) numbers, and their ratios (rational numbers or fractions), were sufficient to describe any quantity. Geometry therefore coupled easily with Pythagorean belief, whose most important tenet was that reality is essentially mathematical and based on whole numbers. Of special relevance was the manipulation of ratios, which at first took place in accordance with rules confirmed by arithmetic. The discovery of surds (the square roots of numbers that are not squares) therefore undermined the Pythagoreans: no longer could $a:b = c:d$ (where a and b, say, are relatively prime) imply that $a = nc$ or $b = nd$, where n is some whole number. According to legend, the Pythagorean discoverer of incommensurable quantities, now known as irrational numbers, was killed by his brethren. But it is hard to keep a secret in science.

The ancient Greeks did not have algebra or Hindu-Arabic numerals. Greek geometry was based almost exclusively on logical reasoning involving abstract diagrams. The discovery of incommensurables, therefore, did more than disturb the Pythagorean notion of the world; it led to an impasse in mathematical reasoning—an impasse that persisted until geometers of Plato's time introduced a definition of proportion (ratio) that accounted for incommensurables. The main mathematicians involved were the Athenian Theaetetus (c. 417–369 BCE), to whom Plato dedicated an entire dialogue, and the great Eudoxus of Cnidus (c. 390–c. 340 BCE), whose treatment of incommensurables survives as Book V of Euclid's *Elements*.

Euclid gave the following simple proof. A square with sides of length 1 unit must, according to the Pythagorean theorem, have a diagonal d that satisfies the equation $d^2 = 1^2 + 1^2 = 2$. Let it be supposed, in accordance with the Pythagorean expectation, that the diagonal can be expressed as the ratio of two integers, say p and q, and that p and q are relatively prime, with $p > q$—in other words, that the ratio has been reduced to its simplest form. Thus $p^2/q^2 = 2$. Then $p^2 = 2q^2$, so p must be an even number, say $2r$. Inserting $2r$ for p in the last equation and simplifying, we obtain $q^2 = 2r^2$, whence q must also be even, which contradicts the assumption that p and q have no common factor other than unity. Hence, no ratio of integers—that is, no "rational number" according to Greek terminology—can express the square root of 2. Lengths such that the squares formed on them are not equal to square numbers (e.g., $\sqrt{2}$, $\sqrt{3}$, $\sqrt{5}$, $\sqrt{6}$,...) were called "irrational numbers."

ISOMETRIC DRAWING

Isometric drawing, which is also known as isometric projection, is a method of graphic representation of three-dimensional objects, used by engineers, technical illustrators, and, occasionally, architects. The technique is intended to combine the illusion of depth, as in a perspective rendering, with the undistorted presentation of the object's principal dimensions, that is, those parallel to a chosen set of three mutually perpendicular coordinate axes.

The isometric is one class of orthographic projections. (In making an orthographic projection, any point in the object is mapped onto the drawing by dropping a perpendicular from that point to the plane of the drawing.) An isometric projection results if the plane is oriented so that it makes equal angles (hence "isometric," or "equal measure") with the three principal planes of the object. Thus, in an isometric drawing of a cube, the three visible faces appear as equilateral parallelograms. That is, while all of the parallel edges of the cube are projected as parallel lines, the horizontal edges are drawn at an angle (usually 30°) from the normal horizontal axes, and the vertical edges, which are parallel to the principal axes, appear in their true proportions.

KÖNIGSBERG BRIDGE PROBLEM

The Königsberg bridge problem is a recreational mathematical puzzle, set in the old Prussian city of Königsberg (now Kaliningrad, Russia), that led to the development of the branches of mathematics known as topology and graph theory. In the early 18th century, the citizens of Königsberg spent their days walking on the intricate arrangement of bridges across the waters of the Pregel (Pregolya) River, which surrounded two central landmasses connected by a bridge. Additionally, the first landmass (an island) was connected by two bridges to the lower bank of the Pregel and also by two bridges to the upper bank, while the other landmass (which split the Pregel into two branches) was connected to the lower bank by one bridge and to the upper bank by one bridge, for a total of seven bridges. According to folklore, the question arose of whether a citizen could take a walk through the town in such a way that each bridge would be crossed exactly once.

In 1735 the Swiss mathematician Leonhard Euler presented a solution to this problem, concluding that such a walk was impossible. To confirm this, suppose that such a walk is possible. In a single encounter with a specific landmass, other than the initial or terminal one, two different bridges must be accounted for: one for entering the landmass and one for leaving it. Thus, each such landmass must serve as an endpoint of a number of bridges equaling twice the number of times it is encountered during the walk. Therefore, each landmass, with the possible exception of the initial and terminal ones if they are not identical, must serve as an endpoint of an even number of bridges. However, for the landmasses of Königsberg, A is an endpoint of five bridges, and B, C, and D are endpoints of three bridges. The walk is therefore impossible.

It would be nearly 150 years before mathematicians would picture the Königsberg bridge problem as a graph consisting of nodes (vertices) representing the landmasses and arcs (edges) representing the bridges. The degree of a vertex of a graph specifies the number of edges incident to it. In modern graph theory, an Eulerian path traverses each edge of a graph once and only once. Thus, Euler's assertion that a graph possessing such a path has at most two vertices of odd degree was the first theorem in graph theory.

Euler described his work as *geometria situs*—the "geometry of position." His work on this problem and some of his later work led directly to the fundamental ideas of combinatorial topology, which 19th-century mathematicians referred to as *analysis situs*—the "analysis of position." Graph theory and topology, both born in the work of Euler, are now major areas of mathematical research.

LINE

Lines are a basic element of Euclidean geometry. Euclid defined a line as an interval between two points and claimed it could be extended indefinitely in either direction. Such an extension in both directions is now thought of as a line, while Euclid's original definition is considered a line segment. A ray is part of a line extending indefinitely from a point on the line in only one direction. In a coordinate system on a plane, a line can be represented by the linear equation $ax + by + c = 0$. This is often written in the slope-intercept form as $y = mx + b$, in which m is the slope and b is the value where the line crosses the y-axis. Because geometrical objects whose edges are line segments are completely understood, mathematicians frequently try to reduce more complex structures into simpler ones made up of connected line segments.

MEASURING THE EARTH, CLASSICAL AND ARABIC

In addition to the attempts of Eratosthenes of Cyrene (c. 276–c. 194 BCE) to measure the Earth, two other early attempts had a lasting historical impact, since they provided values that Christopher Columbus (1451–1506) exploited in selling his project to reach Asia by traveling west from Europe. One was devised by the Greek philosopher Poseidonius (c. 135–c.

51 BCE), the teacher of the great Roman statesman Marcus Tullius Cicero (106–43 BCE). According to Poseidonius, when the star Canopus sets at Rhodes, it appears to be 7.5° above the horizon at Alexandria. (In fact, it is a little over 5°.) Because of the right angles at Rhodes (R) and Alexandria (A) and the parallel lines of sight to Canopus, $\angle RCA$ equals the angular height of Canopus at Alexandria (the errant 7.5°). To obtain the radius $r = CR = CA$, Poseidonius needed the length of the arc RA. It could not be paced out, as travelers from Aswan to Alexandria had done for Eratosthenes' result, because the journey lay over water. Poseidonius could only guess the distance, and his calculation for the size of the Earth was less than three-quarters of what Eratosthenes had found.

The second method, practiced by medieval Arabs, required a free-standing mountain of known height AB. The observer measured $\angle ABH$ between the vertical BA and the line to the horizon BH. Since $\angle BHC$ is a right angle, the Earth's radius $r = CH = AC$ is given by solution of the simple trigonometric equation $\sin(\angle ABH) = {}^{r}/_{(r + AB)}$. The Arab value for the Earth's circumference agreed with the value calculated by Poseidonius—or so Columbus argued, ignoring or forgetting that the Arabs expressed their results in Arab miles, which were longer than the Roman miles with which Poseidonius worked. By claiming that the "best" measurements agreed that the real Earth was three-fourths the size of Eratosthenes' Earth, Columbus reassured his backers that his small wooden ships could survive the journey—he put it at 30 days—to "Cipangu" (Japan).

MEASURING THE EARTH, MODERN

The fitting of lenses to surveying instruments in the 1660s greatly improved the accuracy of the Greek method of measuring the Earth, and this soon became the preferred technique. In its modern form, the method requires the following elements: two stations on the same meridian of longitude, which play the same parts as Aswan and Alexandria in the method of Eratosthenes of Cyrene (c. 276–c. 194 BCE); a precise determination of the angular height of a designated star at the same time from the two stations; and two perfectly level and accurately measured baselines a few kilometres long near each station. What was new 2,000 years after Eratosthenes was the accuracy of the stellar positions and the measured distance between the stations, accomplished through the use of the baselines.

During the 18th century surveyors and astronomers, practicing their updated Greek geodesy in Lapland and Peru, corroborated the conclusion of Isaac Newton (1643–1727), deduced at his desk in Cambridge, England, that the Earth's equatorial axis exceeds its polar axis by a few miles. So precise was the method that subsequent investigation using it revealed that the Earth does not have the shape of an ellipsoid of revolution (an ellipse rotated around one of its axes) but rather has an ineffable shape of its own, now known as the geoid. The method further established the fundamental grids for the mapping of Europe and its colonies. During the French Revolution modernized Greek geodesy was employed to find the equivalent, in the old royal system of measurement, of the new fundamental unit, the standard meter. By definition, the meter was one ten-millionth part of a quarter of the meridian through Paris, making the Earth circumference a nominal 40,000 kilometres.

METRIC SPACE

A metric space is an abstract set with a distance function, called a metric, that specifies a nonnegative distance between any two of its points in such a way that the following properties hold: (1) the distance from the first point to the second equals zero if and only if the points are the same, (2) the distance from the first point to the second equals the distance from the second to the first, and (3) the sum of the distance from the first point to the second and the distance from the second point to a third exceeds or equals the distance from the first to the third. The last of these properties is called the triangle inequality. The French mathematician Maurice Fréchet initiated the study of metric spaces in 1905.

The usual distance function on the real number line is a metric, as is the usual distance function in Euclidean n-dimensional space. There are also more exotic examples of interest to mathematicians. Given any set of points, the discrete metric specifies that the distance from a point to itself equal 0 while the distance between any two distinct points equal 1. The so-called taxicab metric on the Euclidean plane declares the distance from a point (x, y) to a point (z, w) to be $|x - z| + |y - w|$. This "taxicab distance" gives the minimum length of a path from (x, y) to (z, w) constructed from horizontal and vertical line segments. In analysis there are several useful metrics on sets of bounded real-valued continuous or integrable functions.

Thus, a metric generalizes the notion of usual distance to more general settings. Moreover, a metric on a set x determines a collection of open sets, or topology, on x when a subset U of x is declared to be open if and only if for each point p of x there is a positive (possibly very small) distance r such that the set of all points of x of distance less than r from p is completely contained in U. In this way metric spaces provide important examples of topological spaces.

A metric space is said to be complete if every sequence of points in which the terms are eventually pairwise arbitrarily close to each other (a so-called Cauchy sequence) converges to a point in the metric space. The usual metric on the rational numbers is not complete since some Cauchy sequences of rational numbers do not converge to rational numbers. For example, the rational number sequence 3, 3.1, 3.14, 3.141, 3.1415, 3.14159, . . . converges to π, which is not a rational number. However, the usual metric on the real numbers is complete, and, moreover, every real number is the limit of a Cauchy sequence of rational numbers. In this sense, the real numbers form the completion of the rational numbers. The proof of this fact, given in 1914 by the German mathematician Felix Hausdorff, can be generalized to demonstrate that every metric space has such a completion.

PARABOLA

A parabola, one of the conic sections, is an open curve produced by the intersection of a right circular cone and a plane parallel to an element of the cone. As a plane curve, it may be defined as the path (locus) of a point moving so that its distance from a fixed line (the directrix) is equal to its distance from a fixed point (the focus). The vertex of the parabola is the point on the curve that is closest to the directrix. It is equidistant from the directrix and the focus. The vertex and the focus determine a line, perpendicular to the directrix, that is the axis of the parabola. The line through the focus parallel to the directrix is the latus rectum (straight side). The parabola is symmetric about its axis, moving farther from the axis as the curve recedes in the direction away from its vertex. Rotation of a parabola about its axis forms a paraboloid.

The parabola is the path, neglecting air resistance and rotational effects, of a projectile thrown outward into the air. The parabolic shape also is seen in certain bridges, forming arches.

For a parabola the axis of which is the x-axis and with vertex at the origin, the equation is $y^2 = 2px$, in which p is the distance between the directrix and the focus.

PARABOLOID

A paraboloid is an open surface generated by rotating a parabola about its axis. If the axis of the surface is the z axis and the vertex is at the origin, the intersections of the surface with planes parallel to the xz and yz planes are parabolas. The intersections of the surface with planes parallel to and above the xy plane are circles. The general equation for this type of paraboloid is $x^2/a^2 + y^2/b^2 = z$.

If $a = b$, intersections of the surface with planes parallel to and above the xy plane produce circles, and the figure generated is the paraboloid of revolution. If a is not equal to b, intersections with planes parallel to the xy plane are ellipses, and the surface is an elliptical paraboloid.

If the surface of the paraboloid is defined by the equation $x^2/a^2 - y^2/b^2 = z$, cuts parallel to the xz and yz planes produce parabolas of intersection, and cutting planes parallel to xy produce hyperbolas. Such a surface is a hyperbolic paraboloid.

A circular or elliptical paraboloid surface may be used as a parabolic reflector. Applications of this property are used in automobile headlights, solar furnaces, radar, and radio relay stations.

PARALLEL POSTULATE

The parallel postulate is one of the five basic postulates, or axioms, of Euclid underpinning Euclidean geometry. It states that through any given point not on a line there passes exactly one line parallel to that line in the same plane. Unlike Euclid's other four postulates, it never seemed entirely self-evident, as attested by efforts to prove it through the centuries. The uniqueness of Euclidean geometry, and the absolute identification of mathematics with reality, was broken in the 19th century when Nikolay Lobachevsky and János Bolyai (1802–60) independently discovered that altering the parallel postulate resulted in perfectly consistent non-Euclidean geometries.

PENCIL

All the lines in a plane passing through a point, or in three dimensions, all the planes passing through a given line is called a pencil. This line is known as the axis of the pencil. In the duality of solid geometry, the duality being a kind of symmetry between points and planes, the dual of a pencil of planes consists of a line of points. In a plane, in which there is a duality between points and lines, the dual of a line of points is the pencil of lines through a point.

PI

The ratio of the circumference of a circle to its diameter is called pi, or π. The symbol π was devised by British mathematician William Jones in 1706 to represent the ratio and was later popularized by Swiss mathematician Leonhard Euler. Because pi is irrational (not equal to the ratio of any two whole numbers), its digits do not repeat, and an approximation such as 22/7 is often used for everyday calculations. To 39 decimal places, pi is 3.14 1592653589793238462643383279502884197.

The Babylonians (c. 2000 BCE) used 3.125 to approximate pi, a value they obtained by calculating the perimeter of a hexagon inscribed within a circle and assuming that the ratio of the hexagon's perimeter to the circle's circumference was 24/25. The Rhind papyrus (c. 1650 BCE) indicates that ancient Egyptians used a value of 256/81 or about 3.16045. Archimedes (c. 250 BCE) took a major step forward by devising a method to obtain pi to any desired accuracy, given enough patience. By inscribing and circumscribing regular polygons about a circle to obtain upper and lower bounds, he obtained $223/71 < \pi < 22/7$, or an average value of about 3.1418. Archimedes also proved that the ratio of the area of a circle to the square of its radius is the same constant.

Over the ensuing centuries, Chinese, Indian, and Arab mathematicians extended the number of decimal places known through tedious calculations, rather than improvements on Archimedes' method. By the end of the 17th century, however, new methods of mathematical analysis in Europe provided improved ways of calculating pi involving infinite series. For example, Sir Isaac Newton used his binomial theorem to calculate 16 decimal places quickly. Early in the 20th century, the Indian mathematician Srinivasa Ramanujan developed exceptionally efficient ways of calculating pi that were later incorporated into computer algorithms. In the early 21st century, computers calculated pi to more than 22.4 trillion decimal places, as well as its two-| quadrillionth digit.

Pi occurs in various mathematical problems involving the lengths of arcs or other curves, the areas of ellipses, sectors, and other curved surfaces, and the volumes of many solids. It is also used in various formulas of physics and engineering to describe such periodic phenomena as the motion of pendulums, the vibration of strings, and alternating electric currents.

PLATONIC SOLID

Any of the five geometric solids whose faces are all identical, regular polygons meeting at the same three-dimensional angles are known as Platonic solids. Also known as the five regular polyhedra, they consist of the tetrahedron (or pyramid), cube, octahedron, dodecahedron, and icosahedron. Pythagoras (c. 580–c. 500 BCE) probably knew the tetrahedron, cube, and dodecahedron. According to Euclid (fl. c. 300 BCE), the octahedron and icosahedron were first discussed by the Athenian mathematician Theaetetus (c. 417–369 BCE). However, the entire group of regular polyhedra owes its popular name to the great Athenian philosopher Plato (428/427–348/347 BCE), who in his dialogue *Timaeus* associated them with the four basic elements—fire, air, water, and earth—that he supposed to form all matter through their combinations. Plato assigned the tetrahedron, with its sharp points and edges, to the element fire; the cube, with its four-square regularity, to earth; and the other solids concocted from triangles (the octahedron and the icosahedron) to air and water, respectively. The one remaining regular polyhedra, the dodecahedron, with 12 pentagonal faces, Plato assigned to the heavens with its 12 constellations. Because of Plato's systematic development of a theory of the universe based on the five regular polyhedra, they became known as the Platonic solids. Euclid devoted the last book of the *Elements* to the regular polyhedra, which thus serve as so many capstones to his geometry. In particular, his is the first known proof that exactly five regular polyhedra exist. Almost 2,000 years later the astronomer Johannes Kepler (1571–1630) resuscitated the idea of using the Platonic solids to explain the geometry of the universe in his first model of the cosmos. The symmetry, structural integrity, and beauty of these solids have inspired architects, artists, and artisans from ancient Egypt to the present.

POLYGON

A polygon is any closed curve consisting of a set of line segments (sides) connected such that no two segments cross. The simplest polygons are triangles (three sides), quadrilaterals (four sides), and pentagons (five sides). If none of the sides, when extended, intersects the polygon, it is a convex polygon; otherwise it is concave. A polygon with all sides equal is equilateral. One with all interior angles equal is equiangular. Any polygon that is both equilateral and equiangular is a regular polygon (e.g., equilateral triangle, square).

PROJECTION

A projection is a correspondence between the points of a figure and a surface (or line). In plane projections, a series of points on one plane may be projected onto a second plane by choosing any focal point, or origin, and constructing lines from that origin that pass through the points on the first plane and impinge upon the second. This type of mapping is called a central projection. The figures made to correspond by the projection are said to be in perspective, and the image is called a projection of the original figure. If the rays are parallel instead, the projection is likewise called "parallel"; if, in addition, the rays are perpendicular to the plane upon which the original figure is projected, the projection is called "orthogonal." If the two planes are parallel, then the configurations of points will be identical. Otherwise this will not be true.

A second common type of projection is called stereographic projection. It refers to the projection of points from a sphere to a plane. This may be accomplished most simply by choosing a plane through the centre of the sphere and projecting the points on its surface along normals, or perpendicular lines, to that plane. In general, however, projection is possible regardless of the attitude of the plane. Mathematically, it is said that the points on the sphere are mapped onto the plane. If a one-to-one correspondence of points exists, then the map is called conformal.

Projective geometry is the discipline concerned with projections and the properties of projective configurations.

PYTHAGOREAN THEOREM

The Pythagorean theorem is proposition number 47 from Book I of Euclid's *Elements*, the well-known geometric theorem that the sum of the squares on the legs of a right triangle is equal to the square on the hypotenuse (the side opposite the right triangle)—or, in familiar algebraic notation, $a^2 + b^2 = c^2$. Although the theorem has long been associated with the Greek mathematician-philosopher Pythagoras (c. 570–500/490 BCE), it is actually far older. Four Babylonian tablets from circa 1900–1600 BCE indicate some knowledge of the theorem, or at least of special integers known as Pythagorean triples that satisfy it. The theorem is mentioned in the Baudhayana *Sulba-sutra* of India, which was written between 800 and 400 BCE. Nevertheless, the theorem came to be credited to Pythagoras. It is also proposition number 47 from Book I of Euclid's *Elements*.

According to the Syrian historian Iamblichus (c. 250–330 CE), Pythagoras was introduced to mathematics by Thales of Miletus and his pupil Anaximander. In any case, it is known that Pythagoras traveled to Egypt about 535 BCE to further his study, was captured during an invasion in 525 BCE by Cambyses II of Persia and taken to Babylon, and may possibly have visited India before returning to the Mediterranean. Pythagoras soon settled in Croton (now Crotone, Italy) and set up a school, or in modern terms a monastery, where all members took strict vows of secrecy, and all new mathematical results for several centuries were attributed to his name. Thus, not only is the first proof of the theorem not known, there is also some doubt that Pythagoras himself actually proved the theorem that bears his name. Some scholars suggest that the first proof was the one shown in the figure. It was probably independently discovered in several different cultures.

Book I of the *Elements* ends with Euclid's famous "windmill" proof of the Pythagorean theorem. Later in Book VI of the *Elements*, Euclid delivers an even easier demonstration using the proposition that the areas of similar triangles are proportionate to the squares of their corresponding sides. Apparently, Euclid invented the windmill proof so that he could place the Pythagorean theorem as the capstone to Book I. He had not yet demonstrated (as he would in Book V) that line lengths can be manipulated in proportions as if they were commensurable numbers (integers or ratios of integers).

A great many different proofs and extensions of the Pythagorean theorem have been invented. Taking extensions first, Euclid himself showed in a theorem praised in antiquity that any symmetrical regular figures drawn on the sides of a right triangle satisfy the Pythagorean relationship: the figure drawn on the hypotenuse has an area equal to the sum of the areas of the figures drawn on the legs. The semicircles that define Hippocrates of Chios's lunes are examples of such an extension.

In the *Nine Chapters on the Mathematical Procedures* (or *Nine Chapters*), compiled in the 1st century CE in China, several problems are given, along with their solutions, that involve finding the length of one of the sides of a right triangle when given the other two sides. In the *Commentary of Liu Hui*, from the 3rd century, Liu Hui offered a proof of the Pythagorean theorem that called for cutting up the squares on the legs of the right triangle and rearranging them ("tangram style") to correspond to the square on the hypotenuse. Although his original drawing does not survive, the next figure shows a possible reconstruction.

The Pythagorean theorem has fascinated people for nearly 4,000 years; there are now an estimated 367 different proofs, including ones by the Greek mathematician Pappus of Alexandria (flourished c. 320 CE), the Arab mathematician-physician Thābit ibn Qurrah (c. 836–901), the Italian artist-inventor Leonardo da Vinci (1452–1519), and even U.S. President James Garfield (1831–1881).

SPACE-TIME

Space-time is a single concept that recognizes the union of space and time, posited by Albert Einstein in the theories of relativity (1905, 1916).

Common intuition previously supposed no connection between space and time. Physical space was held to be a flat, three-dimensional continuum—that is, an arrangement of all possible point locations—to which Euclidean postulates would apply. To such a spatial manifold, Cartesian coordinates seemed most naturally adapted, and straight lines could be conveniently accommodated. Time was viewed independent of space—as a separate, one-dimensional continuum, completely homogeneous along its infinite extent. Any "now" in time could be regarded as an origin from which to take duration past or future to any other time instant. Within a separately conceived space and time, from the possible states of motion one could not find an absolute state of rest. Uniformly moving spatial coordinate systems attached to uniform time continua represented all unaccelerated motions, the special class of so-called inertial reference frames. The universe according to this convention was called Newtonian.

By use of a four-dimensional space-time continuum, another well-defined flat geometry, the Minkowski universe (after Hermann Minkowski), can be constructed. In that universe, the time coordinate of one coordinate system depends on both the time and space coordinates of another relatively moving system, forming the essential alteration required for Einstein's special theory of relativity. The Minkowski universe, like its predecessor, contains a distinct class of inertial reference frames and is likewise not affected by the presence of matter (masses) within it. Every set of coordinates, or particular space-time event, in such a universe is described as a "here-now" or a world point. Apparent space and time intervals between events depend upon the velocity of the observer, which cannot, in any case, exceed the velocity of light. In every inertial reference frame, all physical laws remain unchanged.

A further alteration of this geometry, locally resembling the Minkowski universe, derives from the use of a four-dimensional continuum containing mass points. This continuum is also non-Euclidean, but it allows for the elimination of gravitation as a dynamical force and is used in Einstein's general theory of relativity (1916). In this general theory, the continuum still consists of world points that may be identified, though non-uniquely, by coordinates. Corresponding to each world point is a coordinate system such that, within the small, local region containing it, the time of special relativity will be approximated. Any succession of these world points, denoting a particle trajectory or light ray path, is known as a world line, or geodesic. Maximum velocities relative to an observer are still defined as the world lines of light flashes, at the constant velocity c.

Whereas the geodesics of a Minkowski continuum (without mass-point accelerations) are straight lines, those of a general relativistic, or Riemannian, universe containing local concentrations of mass are curved; and gravitational fields can be interpreted as manifestations of the space-time curvature. However, one can always find coordinate systems in which, locally, the gravitational field strength is nonexistent. Such a reference frame, affixed to a selected world point, would naturally be in free-fall acceleration near a concentrated mass. Only in this region is the concept well defined—that is, in the neighbourhood of the world point, in a limited region of space, for a limited duration. Its free-fall toward the mass is due either to an externally produced gravitational field or to the equivalent, an intrinsic property of inertial reference frames. Mathematically, gravitational potentials in the Riemannian space can be evaluated by the procedures of tensor analysis to yield a solution of the Einstein gravitational field equations outside the mass points themselves, for any particular distribution of matter.

The first rigorous solution, for a single spherical mass, was carried out by a German astronomer, Karl Schwarzschild (1916). For so-called small masses, the solution does not differ appreciably from that afforded by Newton's gravitational law. However, for "large" masses the radius of space-time curvature may approach or exceed that of the physical object, and the Schwarzschild solution predicts unusual properties. Astronomical observations of dwarf stars eventually led the American physicists J. Robert Oppenheimer and H. Snyder (1939) to postulate super-dense states of matter. These, and other hypothetical conditions of gravitational collapse, were borne out in later discoveries of pulsars and neutron stars. They also have a bearing on black holes thought to exist in interstellar space. Other implications of space-time are important cosmologically and to unified field theory.

Spiral

A spiral is a plane curve that, in general, winds around a point while moving ever farther from the point. Many kinds of spiral are known, the first dating from the days of ancient Greece. The curves are observed in nature, and human beings have used them in machines and in ornament, notably architectural—for example, the whorl in an Ionic capital. The two most famous spirals are described below.

Although Greek mathematician Archimedes did not discover the spiral that bears his name, he did employ it in his *On Spirals* (c. 225 BCE) to square the circle and trisect an angle. The equation of the spiral of Archimedes is $r = a\theta$, in which a is a constant, r is the length of the radius from the centre, or beginning, of the spiral, and θ is the angular position (amount of rotation) of the radius. Like the grooves in a phonograph record, the distance between successive turns of the spiral is a constant—$2\pi a$, if θ is measured in radians.

The equiangular, or logarithmic, spiral was discovered by the French scientist René Descartes in 1638. In 1692 the Swiss mathematician Jakob Bernoulli named it *spira mirabilis* ("miracle spiral") for its mathematical properties; it is carved on his tomb. The general equation of the logarithmic spiral is $r = ae^{\theta \cot b}$, in which r is the radius of each turn of the spiral, a and b are constants that depend on the particular spiral, θ is the angle of rotation as the curve spirals, and e is the base of the natural logarithm. Whereas successive turns of the spiral of Archimedes are equally spaced, the distance between successive turns of the logarithmic spiral increases in a geometric progression (such as 1, 2, 4, 8,...). Among its other interesting properties, every ray from its centre intersects every turn of the spiral at a constant angle (equiangular), represented in the equation by b. Also, for $b = \pi/2$ the radius reduces to the constant a—in other words, to a circle of radius a. This approximate curve is observed in spider webs and, to a greater degree of accuracy, in the chambered mollusk, nautilus, and in certain flowers.

SQUARE

A square is a plane figure with four equal sides and four right (90°) angles. A square is a special kind of rectangle (an equilateral one) and a special kind of parallelogram (an equilateral and equiangular one). A square has four axes of symmetry, and its two finite diagonals (as with any rectangle) are equal. Bisection of a square by a diagonal results in two right triangles. If the length of the side of a square is s, then the area of the square is s^2, or "s squared." From this relation is derived the algebraic use of the term *square*, which denotes the product that results from multiplying any algebraic expression by itself.

THALES' RECTANGLE

Thales of Miletus flourished about 600 BCE and is credited with many of the earliest known geometric proofs. In particular, he has been credited with proving the following five theorems: (1) a circle is bisected by any diameter; (2) the base angles of an isosceles triangle are equal; (3) the opposite ("vertical") angles formed by the intersection of two lines are equal; (4) two triangles are congruent (of equal shape and size) if two angles and a side are equal; and (5) any angle inscribed in a semicircle is a right angle (90°).

Although none of Thales' original proofs survives, the English mathematician Thomas Heath (1861–1940) proposed what is now known as Thales' rectangle would have been consistent with what was known in Thales' era.

Beginning with $\angle ACB$ inscribed in the semicircle with diameter AB, draw the line from C through the corresponding circle's centre O such that it intersects the circle at D. Then complete the quadrilateral by drawing the lines AD and BD. First, note that the lines AO, BO, CO, and DO are equal because each is a radius, r, of the circle. Next, note that the vertical angles formed by the intersection of lines AB and CD form two sets of equal angles, as indicated by the tick marks. Applying a theorem known to Thales, the side-angle-side (SAS) theorem—two triangles are congruent if two sides and the included angle are equal—yields two sets of congruent triangles: $\triangle AOD \cong \triangle BOC$ and $\triangle DOB \cong \triangle COA$. Since the triangles are congruent, their corresponding parts are equal: $\angle ADO = \angle BCO$, $\angle DAO = \angle CBO$, $\angle BDO = \angle ACO$, and so forth. Since all of these triangles are isosceles, their base angles are equal, which means that there are two sets of four angles that are equal, as indicated by the tick marks. Finally, since each angle of the quadrilateral has the same composition, the four quadrilateral angles must be equal—a result that is only possible for a rectangle. Therefore, $\angle ACB = 90°$.

TOPOLOGICAL SPACE

Topological spaces are a generalization of Euclidean spaces in which the idea of closeness, or limits, is described in terms of relationships between sets rather than in terms of distance. Every topological space consists of: (1) a set of points; (2) a class of subsets defined axiomatically as open sets; and (3) the set operations of union and intersection. In addition, the class of open sets in (2) must be defined in such a manner that the intersection of any finite number of open sets is itself open and the union of any, possibly infinite, collection of open sets is likewise open. The concept of limit point is of fundamental importance in topology. A point p is called a limit point of the set S if every open set containing p also contains some point (s) of S (points other than p, should p happen to lie in S). The concept of limit point is so basic to topology that, by itself, it can be used axiomatically to define a topological space by specifying limit points for each set according to rules known as the Kuratowski closure axioms. Any set of objects can be made into a topological space in various ways, but the usefulness of the concept depends on the manner in which the limit points are separated from each other. Most topological spaces that are studied have the Hausdorff property, which states that any two points can be contained in nonoverlapping open sets, guaranteeing that a sequence of points can have no more than one limit point.

Chapter 4

BIOGRAPHIES OF GREAT GEOMETERS

For both practical and aesthetic reasons, the contemplation of geometry has long fascinated and delighted mathematicians and amateurs alike. Indeed, the very notion of mathematics as a logically rigorous endeavour dates back to the ancient Greeks, who first developed the notion of proceeding from a few universally accepted propositions, or axioms, to ever more abstract theorems, which often had no appreciable import beyond the joy derived from their proof. This delight in proofs was carried by Greek contacts (trade and conquest) from the Mediterranean across the Middle East to India, where successive cultures took up the quest for geometric knowledge. The first section of biographies highlights some of the more prominent individuals from this early period.

With the final fall in the 15th century of the Eastern Roman Empire, or Byzantine Empire, knowledge from the rich Greek past and the contributions from the East began to flow into Western Europe. This helped bring an end to the Middle Ages and spawned the Renaissance, which included the discovery of new forms of geometry such as projective geometry and analytic geometry. The second section of biographies highlights some of the more prominent mathematicians from near the end of the Middle Ages to about 1800.

Geometry fell somewhat out of fashion after the discovery of calculus at the end of the 17th century and the growing power of algebra in the following centuries, but soon these disciplines were applied to geometry to create new fields such as algebraic geometry, differential geometry, non-Euclidean geometries, and topology. The third section of biographies highlights some of the more prominent mathematicians since about 1800.

Ancient Greek and Islamic Geometers

Apollonius of Perga

(b. c. 240 BCE, Perga, Pamphylia, Anatolia—
d. c. 190 BCE, Alexandria, Egypt)

Apollonius of Perga was a mathematician, known by his contemporaries as "the Great Geometer," whose treatise *Conics* is one of the greatest scientific works from the ancient world. Most of his other treatises are now lost, although their titles and a general indication of their contents were passed on by later writers, especially Pappus of Alexandria (fl. c. 320 CE). Apollonius's work inspired much of the advancement of geometry in the Islamic world in medieval times, and the rediscovery of his *Conics* in Renaissance Europe formed a good part of the mathematical basis for the scientific revolution.

As a youth, Apollonius studied in Alexandria (under the pupils of Euclid, according to Pappus) and subsequently taught at the university there. He visited both Ephesus and Pergamum, the latter being the capital of a Hellenistic kingdom in western Anatolia, where a university and library similar to the Library of Alexandria had recently been built. In Alexandria he wrote the first cdition of *Conics*, his classic treatise concerning the curves—circle, ellipse, parabola, and hyperbola—that can be generated by intersecting a plane with a cone. He later confessed to his friend Eudemus, whom he had met in Pergamum, that he had written the first version "somewhat too hurriedly." He sent copies of the first three chapters of

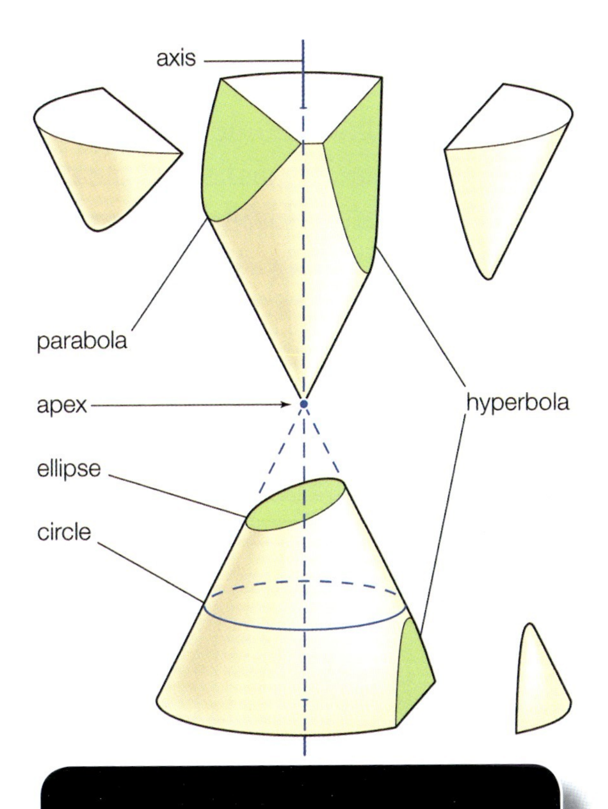

There are three distinct families of conic sections: the ellipse, the parabola, and the hyperbola.

the revised version to Eudemus and, upon Eudemus's death, sent versions of the remaining five books to one Attalus, whom some scholars identify as King Attalus I of Pergamum.

No writings dedicated to conic sections before Apollonius survive, for his *Conics* superseded earlier treatises as surely as Euclid's *Elements* had obliterated earlier works of that genre. Although it is clear that Apollonius made the fullest use of his predecessors' works, such as the treatises of Menaechmus (fl. c. 350 BCE), Aristaeus (fl. c. 320 BCE), Euclid (fl. c. 300 BCE), Conon of Samos (fl. c. 250 BCE), and Nicoteles of Cyrene (fl. c. 250 BCE), he introduced new generality. Whereas his predecessors had used finite right circular cones, Apollonius considered arbitrary (oblique) double cones that extend indefinitely in both directions.

The first four books of the *Conics* survive in the original Greek, the next three only from a 9th-century Arabic translation, and an eighth book is now lost. Books I–IV contain a systematic account of the essential principles of conics and introduce the terms ellipse, parabola, and hyperbola, by which they became known. Although most of Books I–II are based on previous works, a number of theorems in Book III and the greater part of Book IV are new. It is with Books V–VII, however, that Apollonius demonstrates his originality. His genius is most evident in Book V, in which he considers the shortest and the longest straight lines that can be drawn from a given point to points on the curve. (Such considerations, with the introduction of a coordinate system, lead immediately to a complete characterization of the curvature properties of the conics.)

The only other extant work of Apollonius is "Cutting Off of a Ratio," in an Arabic translation. Pappus mentions five additional works, "Cutting Off of an Area" (or "On Spatial Section"), "On Determinate Section," "Tangencies," "Vergings" (or "Inclinations"), and "Plane Loci," and provides valuable information on their contents in Book VII of his *Collection*.

Many of the lost works were known to medieval Islamic mathematicians, however, and it is possible to obtain a further idea of their contents through citations found in the medieval Arabic mathematical literature. For instance, "Tangencies" embraced the following general problem: given three things, each of which may be a point, straight line, or circle, construct a circle tangent to the three. Sometimes known as the problem of Apollonius, the most difficult case arises when the three given things are circles.

Of the other works of Apollonius referred to by ancient writers, one, "On the Burning Mirror," concerned optics. Apollonius demonstrated that parallel light rays striking the interior surface of a spherical mirror would not be reflected to the centre of sphericity, as was previously believed; he also discussed the focal properties of parabolic mirrors. A work titled "On the Cylindrical Helix" is mentioned by Proclus (c. 410–485 CE). According to the mathematician Hypsicles of Alexandria (c. 190–120 BCE), Apollonius also wrote "Comparison of the Dodecahedron and the Icosahedron," on the ratios between both the volumes and the surface areas of these Platonic solids when they are inscribed in the same sphere. According to the mathematician Eutocius of Ascalon (c. 480–540 CE), in Apollonius's work "Quick Delivery," closer limits for the value of π than the 3 10/71 and 3 1/7 of Archimedes (c. 290– 212/211 BCE) were calculated. His "On Unordered Irrationals" extended the theory of irrationals found in Book X of Euclid's *Elements*.

Lastly, from references in Ptolemy's *Almagest*, it is known that Apollonius proved the equivalence of a system of eccentric planetary motion with a special case of epicyclic motion. Of particular interest was his determination of the points where, under general epicyclic motion, a planet appears stationary.

ARCHIMEDES

(B. C. 287 BCE, SYRACUSE, SICILY [NOW IN ITALY]— D. 212/211 BCE, SYRACUSE)

Archimedes was the most famous mathematician and inventor of ancient Greece. Archimedes is especially important for his discovery of the relation between the surface and volume of a sphere and its circumscribing cylinder. He is known for his formulation of a hydrostatic principle (known as Archimedes' principle) and a device for raising water, still used in developing countries, known as the Archimedes screw.

Archimedes probably spent some time in Egypt early in his career, but he resided for most of his life in

Archimedes, oil on canvas by Giuseppe Nogari, 18th century; in the Pushkin Fine Arts Museum, Moscow. © Fine Art Images/age fotostock

Syracuse, the principal Greek city-state in Sicily, where he was on intimate terms with its king, Hieron II. Archimedes published his works in the form of correspondence with the principal mathematicians of his time, including the Alexandrian scholars Conon of Samos and Eratosthenes of Cyrene. He played an important role in the defense of Syracuse against the siege laid by the Romans in 213 BCE by constructing war machines so effective that they long delayed the capture of the city. When Syracuse eventually fell to the Roman general Marcus Claudius Marcellus in the autumn of 212 or spring of 211 BCE, Archimedes was killed in the sack of the city.

According to Plutarch (c. 46–119 CE), Archimedes had so low an opinion of the kind of practical invention at which he excelled and to which he owed his contemporary fame that he left no written work on such subjects. While it is true that—apart from a dubious reference to a treatise, "On Sphere Making"—all of his known works were of a theoretical character, his interest in mechanics nevertheless deeply influenced his mathematical thinking. Not only did he write works on theoretical mechanics and hydrostatics, but his treatise *Method Concerning Mechanical Theorems* shows that he used mechanical reasoning as a heuristic device for the discovery of new mathematical theorems.

There are nine extant treatises by Archimedes in Greek. The principal results in *On the Sphere and Cylinder* (in two books) are that the surface area of any sphere of radius r is four times that of its greatest circle (in modern notation, $S = 4\pi r^2$) and that the volume of a sphere is two-thirds that of the cylinder in which it is inscribed (leading immediately to the formula for the volume, $V = {}^{4}/_{3}\pi r^{3}$). Archimedes was proud enough of the latter discovery to leave instructions for his tomb to be marked with a sphere inscribed in a cylinder. Marcus Tullius Cicero (106–43 BCE) found the tomb, overgrown with vegetation, a century and a half after Archimedes' death.

Measurement of the Circle is a fragment of a longer work in which π (pi), the ratio of the circumference to the diameter of a circle, is shown to lie between the limits of $3\ {}^{10}/_{71}$ and $3\ {}^{1}/_{7}$. Archimedes' approach to determining π, which consists of inscribing and circumscribing regular polygons with a large number of sides, was followed by everyone until the development of infinite series expansions in India during the 15th century and in Europe during the 17th century. That work also contains accurate approximations (expressed as ratios of integers) to the square roots of 3 and several large numbers.

On Conoids and Spheroids deals with determining the volumes of the segments of solids formed by the revolution of a conic section (circle, ellipse, parabola, or hyperbola) about its axis. In modern terms, those are problems of integration. *On Spirals* develops many properties of tangents to, and areas associated with, the spiral of Archimedes—i.e., the locus of a point moving with uniform speed along a straight line that itself is rotating with uniform speed about a fixed point. It was one of only a few curves beyond the straight line and the conic sections known in antiquity.

On the Equilibrium of Planes (or *Centres of Gravity of Planes*; in two books) is mainly concerned with establishing the centres of gravity of various rectilinear plane figures and segments of the parabola and the paraboloid. The first book purports to establish the "law of the lever" (magnitudes balance at distances from the fulcrum in inverse ratio to their weights), and it is mainly on the basis of that treatise that Archimedes has been called the founder of theoretical mechanics. Much of that book, however, is undoubtedly not authentic, consisting as it does of inept later additions or reworkings, and it seems likely that the basic principle of the law of the lever and—possibly—the concept of the centre of gravity were established on a mathematical basis by scholars earlier than Archimedes. His contribution was rather to extend those concepts to conic sections.

Quadrature of the Parabola demonstrates, first by "mechanical" means (as in *Method*, discussed below) and then by conventional geometric methods, that the area of any segment of a parabola is $^4/_3$ of the area of the triangle having the same base and height as that segment. That is, again, a problem in integration.

The Sand-Reckoner is a small treatise that is written for the layman—it is addressed to Gelon, son of Hieron—that nevertheless contains some profoundly original mathematics. Its object is to remedy the inadequacies of the Greek numerical notation system by showing how to express a huge number—the number of grains of sand that it would take to fill the whole of the universe. What Archimedes does, in effect, is to create a place-value system of notation, with a base of 100,000,000. (That was apparently a completely original idea, since he had no knowledge of the contemporary Babylonian place-value system with base 60.) The work is also of interest because it gives the most detailed surviving description of the heliocentric system of Aristarchus of Samos (c. 310–230 BCE) and because it contains an account of an ingenious procedure that Archimedes used to determine the Sun's apparent diameter by observation with an instrument.

Method Concerning Mechanical Theorems describes a process of discovery in mathematics. It is the sole surviving work from antiquity, and one of the few from any period, that deals with this topic. In it Archimedes recounts how he used a "mechanical" method to arrive at some of his key discoveries, including the area of a parabolic segment and the surface area and volume of a sphere. The technique consists of dividing each of two figures into an infinite but equal number of infinitesimally thin strips, then "weighing" each corresponding pair of these strips against each other on a notional balance to obtain the ratio of the two original figures. Archimedes emphasizes that, though useful as a heuristic method, this procedure does not constitute a rigorous proof.

On Floating Bodies (in two books) survives only partly in Greek, the rest in medieval Latin translation from the Greek. It is the first known work on hydrostatics, of which Archimedes is recognized as the founder.

Its purpose is to determine the positions that various solids will assume when floating in a fluid, according to their form and the variation in their specific gravities. In the first book various general principles are established, notably what has come to be known as Archimedes' principle: a solid denser than a fluid will, when immersed in that fluid, be lighter by the weight of the fluid it displaces. The second book is a mathematical tour de force unmatched in antiquity and rarely equaled since. In it Archimedes determines the different positions of stability that a right paraboloid of revolution assumes when floating in a fluid of greater specific gravity, according to geometric and hydrostatic variations.

Archimedes' mathematical proofs and presentation exhibit great boldness and originality of thought on the one hand and extreme rigour on the other, meeting the highest standards of contemporary geometry. While *Method* shows that he arrived at the formulas for the surface area and volume of a sphere by "mechanical" reasoning involving infinitesimals, in his actual proofs of the results in *Sphere* and *Cylinder* he uses only the rigorous methods of successive finite approximation that had been invented by Eudoxus of Cnidus in the 4th century BCE. These methods, of which Archimedes was a master, are the standard procedure in all his works on higher geometry that deal with proving results about areas and volumes. Their mathematical rigour stands in strong contrast to the "proofs" of the first practitioners of integral calculus in the 17th century, when infinitesimals were reintroduced into mathematics. Yet Archimedes' results are no less impressive than theirs. The same freedom from conventional ways of thinking is apparent in the arithmetical field in *Sand-Reckoner*, which shows a deep understanding of the nature of the numerical system.

Given the magnitude and originality of Archimedes' achievement, the influence of his mathematics in antiquity was rather small. Those of his results that could be simply expressed—such as the formulas for the surface area and volume of a sphere—became mathematical commonplaces, and one of the bounds he established for π, 22/7, was adopted as the usual approximation to it in antiquity and the Middle Ages. Nevertheless, his mathematical work was not continued or developed, as far as is known, in any important way in ancient times, despite his hope expressed in *Method* that its publication would enable others to make new discoveries. However, when some of his treatises were translated into Arabic in the late 8th or 9th century, several mathematicians of medieval Islam were inspired to equal or improve on his achievements. This holds particularly in the determination of the volumes of solids of revolution, but his influence is also evident in the determination of centres of gravity and in geometric construction problems. Thus, several meritorious works by medieval Islamic mathematicians were inspired by their study of Archimedes.

The greatest impact of Archimedes' work on later mathematicians came in the 16th and 17th centuries with the printing of texts derived from the Greek, and eventually of the Greek text itself, the *editio princeps*, in Basel in 1544. The Latin translation of many of Archimedes' works by Federico Commandino in 1558 contributed greatly to the spread of knowledge of them, which was reflected in the work of the foremost mathematicians and physicists of the time, including Johannes Kepler (1571–1630) and Galileo Galilei (1564–1642). David Rivault's edition and Latin translation (1615) of the complete works, including the ancient commentaries, was enormously influential in the work of some of the best mathematicians of the 17th century, notably René Descartes (1596–1650) and Pierre de Fermat (1601–1665). Without the background of the rediscovered ancient mathematicians, among whom Archimedes was paramount, the development of mathematics in Europe in the century between 1550 and 1650 is inconceivable. It is unfortunate that *Method* remained unknown to both Arabic and Renaissance mathematicians (it was only rediscovered in the late 19th century), for they might have fulfilled Archimedes' hope that the work would prove useful in the discovery of theorems.

ARCHYTAS OF TARENTUM

(FL. 400–350 BCE, TARENTUM, MAGNA GRAECIA [NOW TARANTO, ITALY])

Archytas of Terentum was a Greek scientist, philosopher, and major Pythagorean mathematician. Plato, a close friend, made use of his work in mathematics, and there is evidence that Euclid borrowed from him for the treatment of number theory in Book VIII of his *Elements*. Archytas was also an influential figure in public affairs, and he served for seven years as commander in chief of his city.

A member of the second generation of followers of Pythagoras, the Greek philosopher who stressed the significance of numbers in explaining all phenomena, Archytas sought to combine empirical observation with Pythagorean theory. In geometry, he solved the problem of doubling the cube by an ingenious construction in solid geometry using the intersection of a cone, a sphere, and a cylinder. (Earlier, Hippocrates of Chios showed that if a cube of side a is given and b and c are line segments such that $a{:}b = b{:}c = c{:}2a$, then a cube of side b has twice the volume, as required. Archytas's construction showed how, given a, to construct the segments b and c with the proper proportions.)

Archytas also applied the theory of proportions to musical harmony. Thus, he showed that if n and $n + 1$ are any two consecutive whole numbers, then there is no rational number b such that $n{:}b = b{:}(n + 1)$; he was thus able to define intervals of pitch in the enharmonic scale in addition to those already known in the chromatic and diatonic scales. Rejecting earlier views that the pitch of notes sounded on a stringed instrument is related to the length or tension of the strings, he correctly showed instead that pitch is related to the movement of vibrating air. However, he incorrectly asserted that the speed at which the vibrations travel to the ear is a factor in determining pitch.

Archytas's reputation as a scientist and mathematician rests on his achievements in geometry, acoustics, and music theory, rather than on his extremely idealistic explanations of human relations and the nature of society according to Pythagorean number theory. Nonmathematical writings usually attributed to him, including a fragment on legal justice, are most likely the work of other authors.

CONON OF SAMOS

(FL. C. 245 BCE, ALEXANDRIA, EGYPT)

Conon of Samos was a mathematician and astronomer whose work on conic sections (curves of the intersections of a right circular cone with a plane) served as the basis for the fourth book of the *Conics* of Apollonius of Perga (c. 262–190 BCE).

From his observations in Italy and Sicily, Conon compiled the *parapegma*, a calendar of meteorological forecasts and of the risings and settings of the stars. He settled in Alexandria, where he served as court astronomer to Ptolemy III Euergetes I (reigned 246–221). When Berenice II, the consort of Ptolemy III, dedicated her hair as an offering in a temple of Aphrodite and the offering disappeared, Conon claimed he could see where it had been placed among the stars in the region of the constellations Boötes, Leo, and Virgo. He named this constellation Coma Berenices ("Berenice's Hair"), thus immortalizing Berenice and further securing his court position.

Conon became a lifelong friend of Archimedes while the latter was studying in Alexandria and later sent him many of his mathematical findings. According to Pappus of Alexandria (flourished c. 320 CE), Conon discovered the Spiral of Archimedes, a curve that Archimedes used extensively in some of his mathematical investigations.

Conon's works included *De astrologia* ("On Astronomy"), in seven books, which according to Seneca contained Egyptian observations of solar eclipses; however, some historians doubt this. He also wrote *Pros Thrasydaion* ("In Reply to Thrasydaeus"), concerning the intersection points of conics with other conics and with circles. None of his works survive.

ERATOSTHENES OF CYRENE
(B. C. 276 BCE, CYRENE, LIBYA—D. C. 194 BCE, ALEXANDRIA, EGYPT)

Eratosthenes of Cyrene was a Greek scientific writer, astronomer, and poet, who made the first measurement of the size of Earth for which any details are known.

At Syene (now Aswān), some 800 km (500 miles) southeast of Alexandria in Egypt, the Sun's rays fall vertically at noon at the summer solstice. Eratosthenes noted that at Alexandria, at the same date and time, sunlight fell at an angle of about 7.2° from the vertical. (Writing before the Greeks adopted the degree, a Babylonian unit of measure, he actually said "a fiftieth of a circle.") He correctly assumed the Sun's distance to be very great. Its rays therefore are practically parallel when they reach Earth. Given an estimate of the distance between the two cities, he was able to calculate the circumference of Earth, obtaining 250,000 stadia. Earlier estimates of the circumference of Earth had been made (for example, Aristotle says that "some mathematicians" had obtained a value of 400,000 stadia), but no details of their methods have survived. An account of Eratosthenes' method is preserved in the Greek astronomer Cleomedes' *Meteora*. The exact length of the units (stadia) he used is doubtful, and the accuracy of his result is therefore uncertain. His measurement of Earth's circumference may have varied by 0.5 to 17 percent from the value accepted by modern astronomers, but it was certainly in the right range. He also measured the degree of obliquity of the ecliptic (in effect, the tilt of Earth's axis) and wrote a treatise on the octaëteris, an eight-year lunar-solar cycle. His only surviving work is Catasterisms, a book about the constellations, which gives a description and story for each constellation, as well as a count of the number of stars contained in it, but the attribution of this work has been doubted by some scholars. His mathematical work is known principally from the writings of the Greek geometer Pappus of Alexandria, and his geographical work from the first two books of the *Geography* of the Greek geographer Strabo.

After study in Alexandria and Athens, Eratosthenes settled in Alexandria about 255 BCE and became director of the great library there. He tried to fix the dates of literary and political events since the siege of Troy. His writings included a poem inspired by astronomy, as well as works on the theatre and on ethics. Eratosthenes was afflicted by blindness in his old age, and he is said to have committed suicide by voluntary starvation.

EUCLID

GREEK EUKLEIDES (FL. C. 300 BCE, ALEXANDRIA, EGYPT)

Euclid was the most prominent mathematician of Greco-Roman antiquity. He is best known for his treatise on geometry, the *Elements*.

Of Euclid's life nothing is known except what the Greek philosopher Proclus (c. 410–485 CE) reports in his "summary" of famous Greek mathematicians. According to him, Euclid taught at Alexandria in the time of Ptolemy I Soter, who reigned over Egypt from 323 to 285 BCE. Medieval translators and editors often confused him with the philosopher Eukleides of Megara, a contemporary of Plato about a century before, and therefore called him Megarensis. Proclus supported his date for Euclid by writing "Ptolemy once asked Euclid if there was not a shorter road to geometry than through the *Elements*, and Euclid replied that there was no royal road to geometry." Today, few historians challenge the consensus that Euclid was older than Archimedes (c. 290/280–212/211 BCE).

Euclid compiled his *Elements* from a number of works of earlier men. Among these are Hippocrates of

Chios (fl. c. 460 BCE), not to be confused with the physician Hippocrates of Cos (c. 460–377 BCE). The latest compiler before Euclid was Theudius, whose textbook was used in the Academy and was probably the one used by Aristotle (384–322 BCE). The older elements were at once superseded by Euclid's and then forgotten. For his subject matter Euclid doubtless drew upon all his predecessors, but it is clear that the whole design of his work was his own, culminating in the construction of the five regular solids, now known as the Platonic solids.

A brief survey of the *Elements* belies a common belief that it concerns only geometry. This misconception may be caused by reading no further than Books I through IV, which cover elementary plane geometry. Euclid understood that building a logical and rigorous geometry (and mathematics) depends on the foundation—a foundation that Euclid began in Book I with 23 definitions (such as "a point is that which has no part" and "a line is a length without breadth"), five unproved assumptions that Euclid called postulates (now known as axioms), and five further unproved assumptions that he called common notions. Book I then proves elementary theorems about triangles and parallelograms and ends with the Pythagorean theorem.

The subject of Book II has been called geometric algebra because it states algebraic identities as theorems about equivalent geometric figures. Book II contains a construction of "the section," the division of a line into two parts such that the ratio of the larger to the smaller segment is equal to the ratio of the original line to the larger segment.

Euclid's Axioms	
1	Given two points there is one straight line that joins them.
2	A straight line segment can be prolonged indefinitely.
3	A circle can be constructed when a point for its centre and a distance for its radius are given.
4	All right angles are equal.
5	If a straight line falling on two straight lines makes the interior angles on the same side less than two right angles, the two straight lines, if produced indefinitely, meet on that side on which the angles are less than the two right angles.
Euclid's Common Notions	
6	Things equal to the same thing are equal.
7	If equals are added to equals, the wholes are equal.
8	If equals are subtracted from equals, the remainders are equal.
9	Things that coincide with one another are equal.
10	The whole is greater than a part.

(This division was renamed the golden section in the Renaissance after artists and architects rediscovered its pleasing proportions.) Book II also generalizes the Pythagorean theorem to arbitrary triangles, a result that is equivalent to the law of cosines. Book III deals with properties of circles and Book IV with the construction of regular polygons, in particular the pentagon.

Book V shifts from plane geometry to expound a general theory of ratios and proportions that is attributed by Proclus (along with Book XII) to Eudoxus of Cnidus (c. 390–350 BCE). While Book V can be read independently of the rest of the *Elements*, its solution to the problem of incommensurables (irrational numbers) is essential to later books. In addition, it formed the foundation for a geometric theory of numbers until an analytic theory developed in the late 19th century. Book VI applies this theory of ratios to plane geometry, mainly triangles and parallelograms, culminating in the "application of areas," a procedure for solving quadratic problems by geometric means.

Books VII–IX contain elements of number theory, where number (*arithmos*) means positive integers greater than 1. Beginning with 22 new definitions—such as unity, even, odd, and prime—these books develop various properties of the positive integers. For instance, Book VII describes a method, *antanaresis* (now known as the Euclidean algorithm), for finding the greatest common divisor of two or more numbers; Book VIII examines numbers in continued proportions, now known as geometric sequences (such as ax, ax^2, ax^3, ax^4...); and Book IX proves that there are an infinite number of primes.

According to Proclus, Books X and XIII incorporate the work of the Pythagorean Theaetetus (c. 417–369

BCE). Book X, which comprises roughly one-fourth of the *Elements*, seems disproportionate to the importance of its classification of incommensurable lines and areas (although study of this book would inspire Johannes Kepler [1571–1630] in his search for a cosmological model). Books XI–XIII examine three-dimensional figures, in Greek stereometria. Book XI concerns the intersections of planes, lines, and parallelepipeds (solids with parallel parallelograms as opposite faces). Book XII applies Eudoxus' method of exhaustion to prove that the areas of circles are to one another as the squares of their diameters and that the volumes of spheres are to one another as the cubes of their diameters. Book XIII culminates with the construction of the five regular Platonic solids (pyramid, cube, octahedron, dodecahedron, icosahedron) in a given sphere.

The unevenness of the several books and the varied mathematical levels may give the impression that Euclid was but an editor of treatises written by other mathematicians. To some extent this is certainly true, although it is probably impossible to figure out which parts are his own and which were adaptations from his predecessors. Euclid's contemporaries considered his work final and authoritative. If more was to be said, it had to be as commentaries to the *Elements*.

The Euclidean corpus falls into two groups: elementary geometry and general mathematics. Although many of Euclid's writings were translated into Arabic in medieval times, works from both groups have vanished.

RENDITIONS OF THE *ELEMENTS*

In ancient times, commentaries were written by Heron of Alexandria (fl. c. 62 CE), Pappus of Alexandria (fl. c. 320 CE), Proclus, and Simplicius of Cilicia (fl. c. 530 CE). The father of Hypatia, Theon of Alexandria (c. 335–405 CE), edited the *Elements* with textual changes and some additions; his version quickly drove other editions out of existence, and it remained the Greek source for all subsequent Arabic and Latin translations until 1808, when an earlier edition was discovered in the Vatican.

The immense impact of the *Elements* on Islamic mathematics is visible through the many translations into Arabic from the 9th century forward, three of which must be mentioned: two by al-Ḥajjāj ibn Yūsuf ibn Maṭar, first for the ʿAbbāsid caliph Hārūn al-Rashīd (ruled 786–809) and again for the caliph al-Maʾmūn (ruled 813–833); and a third by Isḥāq ibn Ḥunayn (died 910), son of Ḥunayn ibn Isḥāq (808–873), which was revised by Thābit ibn Qurrah (c. 836–901) and again by Naṣīr al-Dīn al-Ṭūsī (1201–74). Euclid first became known in Europe through Latin translations of these versions.

The first extant Latin translation of the *Elements* was made about 1120 by Adelard of Bath, who obtained a copy of an Arabic version in Spain, where he traveled while disguised as a Muslim student. Adelard also composed an abridged version and an edition with commentary, thus starting a Euclidean tradition of the greatest importance until the Renaissance unearthed Greek manuscripts. Incontestably the best Latin translation from Arabic was made by Gerard of Cremona (c. 1114–87) from the Isḥāq-Thābit versions.

The first direct translation from the Greek without an Arabic intermediary was made by Bartolomeo Zamberti and published in Vienna in Latin in 1505, and the *editio prin-*

ceps of the Greek text was published in Basel in 1533 by Simon Grynaeus. The first English translation of the *Elements* was by Sir Henry Billingsley in 1570. The impact of this activity on European mathematics cannot be exaggerated; the ideas and methods of Kepler, Pierre de Fermat (1601–65), René Descartes (1596–1650), and Isaac Newton (1642 [Old Style]–1727) were deeply rooted in, and inconceivable without, Euclid's *Elements*.

Extant in the first group is the *Data* (from the first Greek word in the book, dedomena ["given"]), a disparate collection of 94 advanced geometric propositions that all take the following form: given some item or property, then other items or properties are also "given"—that is, they can be determined. Some of the propositions can be viewed as geometry exercises to determine if a figure is constructible by Euclidean means. *On Divisions* (of figures)—restored and edited in 1915 from extant Arabic and Latin versions— deals with problems of dividing a given figure by one or more straight lines into various ratios to one another or to other given areas.

Among Euclid's extant works are the *Optics*, the first Greek treatise on perspective, and the *Phaenomena*, an introduction to mathematical astronomy. Those works are part of a corpus known as "the Little Astronomy" that also includes the *Moving Sphere* by Autolycus of Pitane.

Almost from the time of its writing, the *Elements* exerted a continuous and major influence on human affairs. It was the primary source of geometric reasoning, theorems, and methods at least until the advent of non-Euclidean geometry in the 19th century. It is sometimes said that, other than the Bible, the *Elements* is the most translated, published, and studied of all the books produced in the Western world. Euclid may not have been a first-class mathematician, but he set a standard for deductive reasoning and geometric instruction that persisted, practically unchanged, for more than 2,000 years.

EUDOXUS OF CNIDUS
(B. C. 395–390 BCE, CNIDUS, ASIA MINOR [NOW IN TURKEY]—D. C. 342–337 BCE, CNIDUS)

Eudoxus of Cnidus was a Greek mathematician and astronomer who substantially advanced proportion theory, contributed to the identification of constellations and thus to the development of observational astronomy in the Greek world, and established the first sophisticated, geometrical model of celestial motion. He also wrote on geography and contributed to philosophical discussions in Plato's Academy. Although none of his writings survive, his contributions are known from many discussions throughout antiquity.

According to the 3rd century CE historian Diogenes Laërtius (the source for most biographical details), Eudoxus studied mathematics with Archytas of Tarentum and medicine with Philistion of Locri. At age 23 he attended lectures in Athens, possibly at Plato's Academy (opened c. 387 BCE). After two months he left for Egypt, where he studied with priests for 16 months. Earning his living as a teacher, Eudoxus then returned to Asia Minor, in particular to Cyzicus on the southern shore of the Sea of Marmara, before returning to Athens where he associated with Plato's Academy.

Aristotle preserved Eudoxus's views on metaphysics and ethics. Unlike Plato, Eudoxus held that forms are in perceptible things. He also defined the good as what all things aim for, which he identified with pleasure. He eventually returned to his native Cnidus where he became a legislator and continued his research until his death at age 53. Followers of Eudoxus, including Menaechmus and Callippus, flourished in both Athens and in Cyzicus.

Eudoxus's contributions to the early theory of proportions (equal ratios) forms the basis for the general account of proportions found in Book V of Euclid's *Elements* (c. 300 BCE). Where previous proofs of proportion required separate treatments for lines, surfaces, and solids, Eudoxus provided general proofs. It is unknown, however, how much later mathematicians may have contributed to the form found in the *Elements*. He certainly formulated the bisection principle that given two magnitudes of the same sort one can continuously divide the larger magnitude by at least halves so as to construct a part that is smaller than the smaller magnitude.

Similarly, Eudoxus's theory of incommensurable magnitudes (magnitudes lacking a common measure) and the method of exhaustion (its modern name) influenced Books Xand XII of the *Elements*, respectively.

Archimedes (c. 287–212/211 BCE), in *On the Sphere and Cylinder* and in *Method*, singled out for praise two of Eudoxus's proofs based on the method of exhaustion: that the volumes of pyramids and cones are one-third the volumes of prisms and cylinders, respectively, with the same bases and heights. Various traces suggest that Eudoxus's proof of the latter began by assuming that the cone and cylinder are commensurable, before reducing the case of the cone and cylinder being incommensurable to the commensurable case. Since the modern notion of a real number is analogous to the ancient notion of ratio, this approach may be compared with 19th-century definitions of the real numbers in terms of rational numbers. Eudoxus also proved that the areas of circles are proportional to the squares of their diameters.

Eudoxus is also probably largely responsible for the theory of irrational magnitudes of the form $a \pm b$ (found in the *Elements*, Book X), based on his discovery that the ratios of the side and diagonal of a regular pentagon inscribed in a circle to the diameter of the circle do not fall into the classifications of Theaetetus of Athens (c. 417–369 BCE). According to Eratosthenes of Cyrene (c. 276–194 BCE), Eudoxus also contributed a solution to the problem of doubling the cube—that is, the construction of a cube with twice the volume of a given cube.

Eudoxus is the most innovative Greek mathematician before Archimedes. His work forms the foundation for the most advanced discussions in Euclid's *Elements* and set the stage for Archimedes' study of volumes and surfaces. The theory of proportions is the first completely articulated theory of magnitudes. Although most astronomers seem to have abandoned his astronomical views by the middle of the 2nd century BCE, his principle that every celestial motion is uniform and circular about the centre endured until the time of the 17th-century astronomer Johannes Kepler. Dissatisfaction with Ptolemy's modification of this principle (where he made the centre of the uniform motion distinct from the centre of the circle of motion) motivated many medieval and Renaissance astronomers, including Nicolaus Copernicus (1473–1543).

Heron of Alexandria
(fl. c. 62 CE, Alexandria, Egypt)

Heron of Alexandria was a Greek geometer and inventor whose writings preserved for posterity a knowledge of the mathematics and engineering of Babylonia, ancient Egypt, and the Greco-Roman world.

Heron's most important geometric work, *Metrica*, was lost until 1896. It is a compendium, in three books, of geometric rules and formulas that Heron gathered from a variety of sources, some of them going back to ancient Babylon, on areas and volumes of plane and solid figures. Book I enumerates means of finding the area of various plane figures and the surface areas of common solids. Included is a derivation of Heron's formula (actually, Archimedes' formula) for the area A of a triangle,

$$A = \sqrt{s(s-a)(s-b)(s-c)}$$

in which a, b, and c are the lengths of the sides of the triangle, and s is one-half the triangle's perimeter. Book I also contains an iterative method known by the Babylonians (c. 2000 BCE) for approximating the square root of a number to arbitrary accuracy. (A variation on such an iterative method is frequently employed by computers today.) Book II gives methods for computing volumes of various solids, including the five regular Platonic solids. Book III treats the division of various plane and solid figures into parts according to some given ratio.

Other works on geometry ascribed to Heron are *Geometrica, Stereometrica, Mensurae, Geodaesia, Definitiones*, and *Liber Geëponicus*, which contain problems similar to those in the *Metrica*. However, the first three are certainly not by Heron in their present form, and the sixth consists largely of extracts from the first. Akin to these works is the *Dioptra*, a book on land surveying; it contains a description of the diopter, a surveying instrument used for the same purposes as the modern theodolite. The treatise also contains applications of the diopter to measuring celestial distances and describes a method for finding the distance between Alexandria and Rome from the difference between local times at which a lunar eclipse would be observed at the two cities. It ends with the description of an odometer for measuring the distance a wagon or cart travels. *Catoptrica* ("Reflection") exists only as a Latin translation of a work formerly thought to be a fragment of Ptolemy's Optica. In *Catoptrica* Heron explains the rectilinear propagation of light and the law of reflection.

Of Heron's writings on mechanics, all that remain in Greek are *Pneumatica, Automatopoietica, Belopoeica*, and *Cheirobalistra*. The *Pneumatica*, in two books, describes a menagerie of mechanical devices, or "toys": singing birds, puppets, coin-operated machines, a fire engine, a water organ, and his most famous invention, the aeolipile, the first steam-powered engine. This last device consists of a sphere mounted on a boiler by an axial shaft with two canted nozzles that produce a rotary motion as steam escapes. The *Belopoeica* ("Engines of War") purports to be based on a work by Ctesibius of Alexandria (fl. c. 270 BCE). Heron's *Mechanica*, in three books, survives only in an Arabic translation, somewhat altered. This work is cited by Pappus of Alexandria (fl. 300 CE), as is also the *Baroulcus* ("Methods of Lifting Heavy Weights"). *Mechanica*, which is closely based on the work of Archimedes, presents a wide range of engineering principles, including a theory of motion, a theory of the balance, methods of lifting and transporting heavy objects with mechanical devices, and how to calculate the centre of gravity for various simple shapes. Both *Belopoeica and Mechanica* contain Heron's solution of the problem of two mean proportionals—two quantities, x and y, that satisfy the ratios $a{:}x = x{:}y = y{:}b$, in which a and b are known—which can be used to solve the problem of constructing a cube with double the volume of a given cube. (For the discovery of the mean proportional relationship see Hippocrates of Chios.)

Only fragments of other treatises by Heron remain. One on water clocks is referred to by Pappus and the philosopher Proclus (410–485 CE). Another, a commentary on Euclid's *Elements*, is often quoted in a surviving Arabic work by Abu'l-'Abbās al-Faḍl ibn Ḥātim al-Nayrīzī (c. 865–922).

HIPPIAS OF ELIS
(FL. 5TH CENTURY BCE, ELIS, IN THE PELOPONNESE, GREECE)

Sophist philosopher Hippias of Elis contributed significantly to mathematics by discovering the quadratrix, a special curve he may have used to trisect an angle.

A man of great versatility, with an assurance characteristic of the later Sophists, Hippias lectured on poetry, grammar, history, politics, archaeology, mathematics, and astronomy. His vast literary output included elegies and tragedies besides technical treatises in prose. He is credited with an excellent work on Homer, collections of Greek and foreign literature, and archaeological treatises; but nothing remains except a few fragments. He is depicted in Plato's Protagoras, and two of Plato's minor dialogues are named after him.

HIPPOCRATES OF CHIOS
(FL. C. 440 BCE)

Greek geometer Hippocrates of Chios compiled the first known work on the elements of geometry nearly a century before Euclid. Although the work is no longer extant, Euclid may have used it as a model for his *Elements*.

According to tradition, Hippocrates was a merchant whose goods had been captured by pirates. He went to Athens to prosecute them but met with little success in recovering his property. He remained in Athens, however, where he attended lectures on mathematics and finally took up teaching geometry to support himself. Aristotle (384–322 BCE) recounts a different story, claiming that Hippocrates was cheated by customs officers in Byzantium. He purportedly told this story to show that, although Hippocrates was a good geometer, he was incompetent to handle the ordinary affairs of life.

Hippocrates' *Elements* is known only through references made in the works of later commentators, especially the Greek philosophers Proclus (c. 410–485 CE) and Simplicius of Cilicia (fl. c. 530 CE). In his attempts to square the circle, Hippocrates was able to find the areas of certain lunes, or crescent-shaped figures contained between two intersecting circles. He based this work upon the theorem that the areas of two circles have the same ratio as the squares of their radii. A summary of these quadratures of lunes, written by Eudemus of Rhodes (c. 335 BCE), with elaborate proofs, has been preserved by Simplicius.

The third of the achievements attributed to Hippocrates was the discovery that, given a cube of side a, a cube with double its volume can be constructed if two mean proportionals, x and y, can be determined such that $a:x = x:y = y:2a$. It is also generally thought that

Hippocrates introduced the tactic of reducing a complex problem to a more tractable or simpler problem. His reduction of the problem of "doubling the cube" (a three-dimensional quantity) to finding two lengths (one dimensional quantities) certainly fits this description.

MENAECHMUS

(B. C. 380 BCE, ALOPECONNESUS, ASIA MINOR [NOW TURKEY]—D. C. 320 BCE, CYZICUS? [MODERN KAPIDAĞI YARIMADASI, TURKEY])

Menaechmus was a Greek mathematician and friend of Plato, who is credited with discovering the conic sections. Menaechmus's credit for discovering that the ellipse, parabola, and hyperbola are sections of a cone—produced by the intersection of a plane with the surface of a cone—derives from an epigram of Eratosthenes of Cyrene (c. 276–194 BCE) that refers to cutting the cone "in the triads of Menaechmus." Eutocius of Ascalon (fl. 520 CE) recounts two of Menaechmus's solutions to the problem of constructing a cube with double the volume of a given cube of side a. Menaechmus's solutions use properties of the parabola and hyperbola to produce line segments x and y such that the following continued proportion holds: $a:x = x:y = y:2a$. (Roughly 100 years earlier, Hippocrates of Chios reduced the problem of "doubling the cube" of side a to finding x and y that satisfy this continued proportion.)

According to the philosopher Proclus (c. 410–485), Menaechmus's brother Dinostratus gained fame as a mathematician for discovering how the trisectrix, a curve first invented for trisecting the angle, could be used to construct a square equal in area to a given circle.

Omar Khayyam

(b. May 18, 1048 CE, Neyshābūr [also spelled Nīshāpūr], Khorāsān [now Iran]—
d. December 4, 1131, Neyshābūr)

Omar Khayyam was a Persian mathematician, astronomer, and poet, renowned in his own country and time for his scientific achievements but chiefly known to English-speaking readers through the translation of a collection of his *robāʿīy āt* ("quatrains") in *The Rubáiyát of Omar Khayyám* (1859), by the English writer Edward FitzGerald.

His name, Khayyam, ("Tentmaker") may have been derived from his father's trade. He received a good education in the sciences and philosophy in his native Neyshābūr before traveling to Samarkand (now in Uzbekistan), where he completed the algebra treatise, *Risālah fiʾl-barāhīn ʿalā masāʾil al-jabr waʾl-muq ābalah* ("Treatise on Demonstration of Problems of Algebra"), on which his mathematical reputation principally rests. In this treatise he gave a systematic discussion of the solution of cubic equations by means of intersecting conic sections. Perhaps it was in the context of this work that he discovered how to extend Abu al-Wafā's results on the extraction of cube and fourth roots to the extraction of *n*th roots of numbers for arbitrary whole numbers *n*.

He made such a name for himself that the Seljuq sultan Malik-Shāh invited him to Eṣfahān to undertake the astronomical observations necessary for the reform of the calendar. To accomplish this an observatory was built there, and a new calendar was produced, known as the Jalālī calendar. Based on making 8 of every 33 years leap years, it was more accurate than the present Gregorian calendar, and it was adopted in 1075 by Malik-Shāh. In Eṣfahān he also produced fundamental critiques of Euclid's theory of parallels as well as his theory of proportion. In connection with the former his ideas eventually made their way to Europe, where they influenced the English mathematician John Wallis (1616–1703); in connection with the latter he argued for the important idea of enlarging the notion of number to include ratios of magnitudes (and hence such irrational numbers as $\sqrt{2}$ and π).

His years in Eṣfahān were very productive ones, but after the death of his patron in 1092 the sultan's widow turned against him, and soon thereafter Omar went on a pilgrimage to Mecca. He then returned to Neyshābūr where he taught and served the court as an astrologer. Philosophy, jurisprudence, history, mathematics, medicine, and astronomy are among the subjects mastered by this brilliant man.

Omar's fame in the West rests upon the collection of *robāʿīyāt*, or "quatrains," attributed to him. (A quatrain is a piece of verse complete in four lines, usually rhyming *aaaa* or *aaba*; it is close in style and spirit to the epigram.) Omar's poems had attracted comparatively little attention until they inspired FitzGerald to write his celebrated *The Rubáiyát of Omar Khayyám*, containing such now-famous phrases as "A Jug of Wine, a Loaf of Bread—and Thou," "Take the Cash, and let the Credit go," and "The Flower that once has blown forever dies." These quatrains have been translated into almost every major language and are largely responsible for colouring European ideas about Persian poetry. Some scholars have doubted that Omar wrote poetry. His contemporaries took no notice of his verse, and not until two centuries after his death did a few quatrains appear under his name. Even then, the verses were mostly used as quotations against particular views ostensibly held by Omar, leading some scholars to suspect that they may have been invented and attributed to Omar because of his scholarly reputation.

Each of Omar's quatrains forms a complete poem in itself. It was FitzGerald who conceived the idea of combining a series of these *robāʿīyāt* into a continuous elegy that had an intellectual unity and consistency. FitzGerald's ingenious and felicitous paraphrasing gave his translations a memorable verve and succinctness. They are, however, extremely free translations, and more recently several more faithful renderings of the quatrains have been published.

The verses translated by FitzGerald and others reveal a man of deep thought, troubled by the questions of the nature of reality and the eternal, the impermanence and uncertainty of life, and man's relationship to God. The writer doubts the existence of divine providence and the afterlife, derides religious certainty, and feels keenly man's frailty and ignorance.

Finding no acceptable answers to his perplexities, he chooses to put his faith instead in a joyful appreciation of the fleeting and sensuous beauties of the material world. The idyllic nature of the modest pleasures he celebrates, however, cannot dispel his honest and straightforward brooding over fundamental metaphysical questions.

PAPPUS OF ALEXANDRIA
(FL. C. 320 CE)

Pappus of Alexandria was the most important mathematical author writing in Greek during the later Roman Empire, known for his *Synagoge* ("Collection"), a voluminous account of the most important work done in ancient Greek mathematics. Other than that he was born at Alexandria in Egypt and that his career coincided with the first three decades of the 4th century CE, little is known about his life. Judging by the style of his writings, he was primarily a teacher of mathematics. Pappus seldom claimed to present original discoveries, but he had an eye for interesting material in his predecessors' writings, many of which have not survived outside of his work. As a source of information concerning the history of Greek mathematics, he has few rivals.

Pappus wrote several works, including commentaries on Ptolemy's *Almagest* and on the treatment of irrational magnitudes in Euclid's *Elements*. His principal work, however, was the *Synagoge* (c. 340), a composition in at least eight books (corresponding to the individual rolls of papyrus on which it was originally written). The only Greek copy of the *Synagoge* to pass through the Middle Ages lost several pages at both the beginning and the end; thus, only Books 3 through 7 and portions of Books 2 and 8 have survived. A complete version of Book 8 does survive, however, in an Arabic translation. Book 1 is entirely lost, along with information on its contents. The *Synagoge* seems to have been assembled in a haphazard way from independent shorter writings of Pappus. Nevertheless, such a range of topics is covered that the *Synagoge* has with some justice been described as a mathematical encyclopedia.

The *Synagoge* deals with an astonishing range of mathematical topics; its richest parts, however, concern geometry and draw on works from the 3rd century BCE, the so-called Golden Age of Greek mathematics. Book 2 addresses a problem in recreational mathematics: given that each letter of the Greek alphabet also serves as a numeral (e.g., $\alpha = 1$, $\beta = 2$, $\iota = 10$), how can one calculate and name the number formed by multiplying together all the letters in a line of poetry. Book 3 contains a series of solutions to the famous problem of constructing a cube having twice the volume of a given cube, a task that cannot be performed using only the ruler-and-compass methods of Euclid's *Elements*. Book 4 concerns the properties of several varieties of spirals and other curved lines and demonstrates how they can be used to solve another classical problem, the division of an angle into an arbitrary number of equal parts. Book 5, in the course of a treatment of polygons and polyhedra, describes Archimedes' discovery of the semiregular polyhedra (solid geometric shapes whose faces are not all identical regular polygons). Book 6 is a student's guide to several texts, mostly from the time of Euclid, on mathematical astronomy. Book 8 is about applications of geometry in mechanics; the topics include geometric constructions made under restrictive conditions, for example, using a "rusty" compass stuck at a fixed opening.

The longest part of the *Synagoge*, Book 7, is Pappus's commentary on a group of geometry books by Euclid, Apollonius of Perga, Eratosthenes of Cyrene, and Aristaeus, collectively referred to as the "Treasury of Analysis." "Analysis" was a method used in Greek geometry for establishing the possibility of constructing a particular geometric object from a set of given objects. The analytic proof involved demonstrating a relationship between the sought object and the given ones such that one was assured of the existence of a sequence of basic constructions leading from the known to the unknown, rather as in algebra. The books of the "Treasury," according to Pappus, provided the equipment for performing analysis. With three exceptions the books are lost, and hence the information that Pappus gives concerning them is invaluable.

Pappus's *Synagoge* first became widely known among European mathematicians after 1588, when a posthumous Latin translation by Federico Commandino was printed in Italy. For more than a century afterward, Pappus's accounts of geometric principles and methods stimulated new mathematical research, and his influence is conspicuous in the work of René Descartes (1596–1650), Pierre de Fermat (1601–1665), and Isaac Newton (1642 [Old Style]–1727), among many others. As late as the 19th century, his commentary on Euclid's lost *Porisms* in Book 7 was a subject of living interest for Jean-Victor Poncelet (1788–1867) and Michel Chasles (1793–1880) in their development of projective geometry.

PYTHAGORAS

(B. C. 580 BCE, SAMOS, IONIA [NOW IN GREECE]—D. C. 500 BCE, METAPONTUM, LUCANIA [NOW IN ITALY])

Pythagoras was a Greek philosopher, mathematician, and founder of the Pythagorean brotherhood that, although religious in nature, formulated principles that influenced the thought of Plato and Aristotle and contributed to the development of mathematics and Western rational philosophy.

Pythagoras migrated to southern Italy about 532 BCE, apparently to escape Samos's tyrannical rule, and established his ethico-political academy at Croton (now Crotone, Italy).

Pythagoras demonstrating his Pythagorean theorem in the sand using a stick.

It is difficult to distinguish Pythagoras's teachings from those of his disciples. None of his writings have survived, and Pythagoreans invariably supported their doctrines by indiscriminately citing their master's authority. Pythagoras, however, is generally credited with the theory of the functional significance of numbers in the objective world and in music. Other discoveries often attributed to him (e.g., the incommensurability of the side and diagonal of a square, and the Pythagorean theorem for right triangles) were probably developed only later by the Pythagorean school. More probably the bulk of the intellectual tradition originating with Pythagoras himself belongs to mystical wisdom rather than to scientific scholarship.

Thales of Miletus
(fl. 6th century BCE)

Philosopher Thales of Miletus was renowned as one of the legendary Seven Wise Men, or Sophoi, of antiquity. He is remembered primarily for his cosmology based on water as the essence of all matter, with the Earth a flat disk floating on a vast sea. The Greek historian Diogenes *Laërtius* (fl. 3rd century CE), quoting Apollodorus of Athens (fl. 140 BCE), placed the birth of Thales during the 35th Olympiad (apparently a transcription error; it should read the 39th Olympiad, c. 624 BCE) and his death in the 58th Olympiad (548–545 BCE) at the age of 78.

No writings by Thales survive, and no contemporary sources exist. Thus, his achievements are difficult to assess. Inclusion of his name in the canon of the legendary Seven Wise Men led to his idealization, and numerous acts and sayings, many of them no doubt spurious, were attributed to him, such as "Know thyself" and "Nothing in excess." According to the historian Herodotus (c. 484–c. 425 BCE), Thales was a practical statesman who advocated the federation of the Ionian cities of the Aegean region. The poet-scholar Callimachus (c. 305–c. 240 BCE) recorded a traditional belief that Thales advised navigators to steer by the Little Bear (Ursa Minor) rather than by the Great Bear (Ursa Major), both prominent constellations in the Northern Hemisphere. He is also said to have used his knowledge of geometry to measure the Egyptian pyramids and to calculate the distance from shore of ships at sea. Although such stories are probably apocryphal, they illustrate Thales' reputation. The poet-philosopher Xenophanes (c. 560–c. 478 BCE) claimed that Thales predicted the solar eclipse that stopped the battle between King Alyattes of Lydia (reigned c. 610–c. 560 BCE) and King Cyaxares of Media (reigned 625–585 BCE), evidently on May 28, 585. Modern scholars believe, however, that he could not possibly have had the knowledge to predict accurately either the locality or the character of an eclipse. Thus, his feat was apparently isolated and only approximate. Herodotus spoke of his foretelling the year only. That the eclipse was nearly total and occurred during a crucial battle contributed considerably to his exaggerated reputation as an astronomer.

Thales has been credited with the discovery of five geometric theorems: (1) that a circle is bisected by its diameter, (2) that angles in a triangle opposite two sides of equal length are equal, (3) that opposite angles formed by intersecting straight lines are equal, (4) that the angle inscribed inside a semicircle is a right angle, and (5) that a triangle is determined if its base and the two angles at the base are given. His mathematical achievements are difficult to assess, however, because of the ancient practice of crediting particular discoveries to men with a general reputation for wisdom.

The claim that Thales was the founder of European philosophy rests primarily on Aristotle (384–322 BCE), who wrote that Thales was the first to suggest a single material substratum for the universe—namely, water, or moisture. A likely consideration in this choice was the seeming motion that water exhibits, as seen in its ability to become vapour. For what changes or moves itself was thought by the Greeks to be close to life itself, and to Thales the entire universe was a living organism, nourished by exhalations from water.

Thales' significance lies less in his choice of water as the essential substance than in his attempt to explain nature by the simplification of phenomena and in his search for causes within nature itself rather than in the caprices of anthropomorphic gods. Like his successors the philosophers Anaximander (610–546/545 BCE) and Anaximenes of Miletus (fl. c. 545 BCE), Thales is important in bridging the worlds of myth and reason.

THEAETETUS
(B. C. 417 BCE, ATHENS [GREECE]—D. 369 BCE, ATHENS)

Athenian mathematician Theaetetus had a significant influence on the development of Greek geometry.

Theaetetus was a disciple of Socrates and studied with Theodorus of Cyrene. He taught at some time in Heraclea (located in present-day southern Italy). Plato made Theaetetus the chief subject of two dialogues—Theaetetōs (Theaetetus) and Sophistēs (Sophist)—the former being the major source of information about Theaetetus's life, including his death in a battle between Athens and Corinth in 369 BCE.

Theaetetus made important contributions to the mathematics that Euclid (fl. c. 300 BCE) eventually collected and systematized in his *Elements*. A key area of Theaetetus's work was on incommensurables (which correspond to irrational numbers in modern mathematics), in which he extended the work of Theodorus by devising the basic classification of incommensurable magnitudes into different types that is found in Book X of the *Elements*. He also discovered methods of inscribing in a sphere the five Platonic solids (tetrahedron, cube, octahedron, dodecahedron, and icosahedron), the subject of Book XII of the *Elements*. Finally, he may be the author of a general theory of proportion that was formulated after the numerically based theory of the Pythagoreans (fl. 5th century BCE) yet before that of Eudoxus of Cnidus (c. 400–350 BCE) as described in Book V of the *Elements*.

Pre-Modern (Pre-1800) Geometers

Bonaventura Cavalieri

(b. 1598, Milan [Italy]—d. Nov. 30, 1647, Bologna, Papal States)

Italian mathematician Bonaventura Cavalieri made developments in geometry that were precursors to integral calculus.

As a boy Cavalieri joined the Jesuati, a religious order (sometimes called "Apostolic Clerics of St. Jerome") that followed the rule of St. Augustine and was suppressed in 1668 by Pope Clement IX. Euclid's works stimulated his interest in mathematics, and, after he met Galileo, Cavalieri considered himself a disciple of that great astronomer.

By 1629, when he was appointed professor of mathematics of the University of Bologna, Cavalieri had completely developed his method of indivisibles, a means of determining the size of geometric figures similar to the methods of integral calculus. He delayed publishing his results for six years out of deference to Galileo, who planned a similar work. Cavalieri's work appeared in 1635 and was entitled *Geometria Indivisibilibus Continuorum Nova Quadam Ratione Promota* ("A Certain Method for the Development of a New Geometry of Continuous Indivisibles"). As stated in his *Geometria*, the method of indivisibles was unsatisfactory and fell under heavy criticism, notably from the contemporary Swiss mathematician Paul Guldin. In reply to this criticism, Cavalieri wrote *Exercitationes Geometricae Sex* (1647; "Six Geometrical Exercises"), stating the principle in the more satisfactory form that was widely employed by mathematicians during the 17th century.

Cavalieri was largely responsible for introducing the use of logarithms as a computational tool in Italy through his book *Directorium Generale Uranometricum* (1632; "A General Directory of Uranometry"). His other works include *Lo specchio ustorio ouero trattato delle settioni coniche* (1632; "The Burning Glass; or, A Treatise on Conic Sections") and *Trigonometria plana et sphaerica, linearis et logarithmica* (1643; "Plane, Spherical, Linear, and Logarithmic Trigonometry").

GIOVANNI CEVA

(B. SEPT. 1, 1647, MILAN [ITALY]—
D. MAY 13, 1734, MANTUA [ITALY])

Giovanni Ceva was an Italian mathematician, physicist, and hydraulic engineer best known for the geometric theorem bearing his name concerning straight lines that intersect at a common point when drawn through the vertices of a triangle.

Most details of Ceva's early life are known only through his correspondence and the prefaces to some of his works. He was educated in a Jesuit college in Milan and then at the University of Pisa, where the work of Galileo Galilei (1564–1642) and his followers on geometry and mechanics exerted a great influence on his education and research interests. He may have taught in Pisa during the time when he produced his first major work, *De lineis rectis* (1678; "Concerning Straight Lines"). In this work Ceva proved many geometrical propositions using the properties of the figures' centres of gravity. This work also contains his rediscovery of a version of a theorem of Menelaus of Alexandria (c. 70–130 CE): Given any triangle ABC, with points R, S, T on sides AB, BC, and AC, respectively, the line segments CR, AS, and BT intersect in a single point if and only if

$$({}^{AR}/_{RB})({}^{BS}/_{SC})({}^{CT}/_{TA}) = 1.$$

During this period he was appointed auditor and commissioner to the duke of Mantua, in which position he administered Mantua's economy. He also wrote the four-volume *Opuscula mathematica* (1682; "Mathematical Essays"), an investigation of forces (including the resultant of many different forces and the parallelogram of forces), pendulum motion, and the behaviour of bodies in flowing water.

By 1684 Ceva was appointed mathematician and superintendent of the waters of the Duchy of Mantua. (Although Mantua was annexed by Austria in 1707, Ceva retained this post for the rest of his life.) Having obtained a secure appointment, Ceva soon married, in January 1685, and a daughter, the first of seven children, was born to him in 1687.

Among the works Ceva produced after moving to Mantua are *Geometria motus* (1692; "The Geometry of Motion"), in which he applied geometry to the study of motion; *De re nummaria* (1711; "Concerning Money Matters"), one of the first works in mathematical economics to examine the conditions for equilibrium in a monetary system; and *Opus hydrostaticum* (1728; "Hydrostatics"), on hydraulics.

GIRARD DESARGUES

(B. FEB. 21, 1591, LYON, FRANCE—D. OCT. 1661, FRANCE)

French mathematician Girard Desargues figures prominently in the history of projective geometry. Desargues's work was well known by his contemporaries, but half a century after his death he was forgotten. His work was rediscovered at the beginning of the 19th century, and one of his results became known as Desargues's theorem.

Not much is known about Desargues's early life, which he spent in Lyon where his father worked for the local diocese. In 1626 Desargues proposed a water project to the municipality of Paris, and by 1630 he had become associated with a group of Parisian mathematicians gathered around Father Marin Mersenne. In 1635 Mersenne formed the informal, private Académie Parisienne, whose meetings Desargues attended. Through Mersenne, Desargues had contact with most of the leading French mathematicians of his day. Two of the most prominent, René Descartes and Pierre de Fermat, valued his scientific views. It is generally presumed that Desargues worked as an engineer until he took up architecture about 1645. He lived in Lyon again from about 1649 to 1657 before returning to Paris for the remainder of his life.

In 1636 Desargues published *Exemple de l'une des manières universelles du S.G.D.L. touchant la pratique de la perspective* ("Example of a Universal Method by Sieur Girard Desargues Lyonnais Concerning the Practice of Perspective"), in which he presented a geometric method for constructing perspective images of objects. The painter Laurent de La Hire and the engraver Abraham Bosse found Desargues's method attractive. Bosse, who taught perspective constructions based on Desargues's method at the Royal Academy of Painting and Sculpture in Paris, published a more accessible presentation of this method in *Manière universelle de Mr. Desargues pour pratiquer la perspective* (1648; "Mr. Desargues's Universal Method of Practising Perspective"). In addition this book contains what is now known as Desargues's theorem. Desargues also published a primer on music notation, a technique for stonecutting, and a guide for the construction of sundials.

Desargues's most important work, *Brouillon project d'une atteinte aux événements des rencontres d'un cône avec un plan* (1639; "Rough Draft of Attaining the Outcome of Intersecting a Cone with a Plane"), treats the theory of conic sections in a projective manner. In this very theoretical work Desargues revised parts of the *Conics* by Apollonius of Perga (c. 262–190 BCE). Regardless of its theoretical character, Desargues claimed that it was of use for artisans. This statement misled later historians into seeing a strong connection between his perspective method and his treatment of conic sections. Both disciplines deal with central projections but are otherwise rather different. It is likely, however, that one of Desargues's projective ideas—the concept of points at infinity—came from his theoretical analysis of perspective.

In the 17th century Desargues's new approach to geometry—studying figures through their projections—was appreciated by a few gifted mathematicians, such as Blaise Pascal and Gottfried Wilhelm Leibniz, but it did not become influential. Descartes's algebraic way of treating geometrical problems—published in *Discours de la méthode* (1637; "Discourse on Method")—came to dominate geometrical thinking and Desargues's ideas were forgotten. His *Brouillon project* became known again only after 1822, when Jean-Victor Poncelet drew attention to the fact that in developing projective geometry (which happened while he was a prisoner of war in Russia, 1812–14) he had been preceded—though not inspired—by Desargues in certain aspects.

René Descartes

(b. March 31, 1596, La Haye, Touraine, France—
d. Feb. 11, 1650, Stockholm, Swed.)

René Descartes was a French mathematician, scientist, and philosopher. Because he was one of the first to abandon scholastic Aristotelianism, because he formulated the first modern version of mind-body dualism (from which stems the mind-body problem) and because he promoted the development of a new science grounded in observation and experiment, he has been called the father of modern philosophy. Applying an original system of methodical doubt, he dismissed apparent knowledge derived from authority, the senses, and reason and erected new epistemic foundations on the basis of the intuition that, when he is thinking, he exists; this he expressed in the dictum "I think, therefore I am" (best known in its Latin formulation, "Cogito, ergo sum," though originally written in French, "Je pense, donc je suis"). He developed a metaphysical dualism that distinguishes radically between mind, the essence of which is thinking, and matter, the essence of which is extension in three dimensions. Descartes's metaphysics is rationalist, based on the postulation of innate ideas of mind, matter, and God, but his physics and physiology, based on sensory experience, are mechanistic and empiricist.

Descartes spent the period 1619 to 1628 traveling in northern and southern Europe, where, as he later explained, he studied "the book of the world." While in Bohemia in 1619, he invented analytic geometry, a method of solving geometric problems algebraically and algebraic problems geometrically. He also devised a universal method of deductive reasoning, based on mathematics, that is applicable to all the sciences. This method, which he later formulated in *Discourse on Method* (1637) and *Rules for the Direction of the Mind* (written by 1628 but not published until 1701), consists of four rules: (1) accept nothing as true that is not self-evident, (2) divide problems into their simplest parts, (3) solve problems by proceeding from simple to complex, and (4) recheck the reasoning. These rules are a direct application of mathematical procedures. In addition, Descartes insisted that all key notions and the limits of each problem must be clearly defined.

LEONHARD EULER

(B. APRIL 15, 1707, BASEL, SWITZ.—D. SEPT. 18, 1783, ST. PETERSBURG, RUSSIA)

Leonhard Euler was a Swiss mathematician and physicist, and one of the founders of pure mathematics. He not only made decisive and formative contributions to the subjects of geometry, calculus, mechanics, and number theory but also developed methods for solving problems in observational astronomy and demonstrated useful applications of mathematics in technology and public affairs.

Euler's mathematical ability earned him the esteem of Johann Bernoulli, one of the first mathematicians in Europe at that time, and of his sons Daniel and Nicolas. In 1727 he moved to St. Petersburg, where he became an associate of the St. Petersburg Academy of Sciences and in 1733 succeeded Daniel Bernoulli to the chair of mathematics.

By means of his numerous books and memoirs that he submitted to the academy, Euler carried integral calculus to a higher degree of perfection, developed the theory of trigonometric and logarithmic functions, reduced analytical operations to a greater simplicity, and threw new light on nearly all parts of pure mathematics. Overtaxing himself, Euler in 1735 lost the sight of one eye. Then, invited by Frederick the Great in 1741, he became a member of the Berlin Academy, where for 25 years he produced a steady stream of publications, many of which he contributed to the St. Petersburg Academy, which granted him a pension.

In 1748, in his *Introductio in analysin infinitorum*, he developed the concept of function in mathematical analysis, through which variables are related to each other and in which he advanced the use of infinitesimals and infinite quantities. He did for modern analytic geometry and trigonometry what the *Elements* of Euclid had done for ancient geometry, and the resulting tendency to render mathematics and physics in arithmetical terms has continued ever since. He is known for familiar results in elementary geometry—for example, the Euler line through the orthocentre (the intersection of the altitudes in a triangle), the circumcentre (the centre of the circumscribed circle of a triangle), and the barycentre (the "centre of gravity," or centroid) of a triangle. He was responsible for treating trigonometric functions—i.e., the relationship of an angle to two sides of a triangle—as numerical ratios rather than as lengths of geometric lines and for relating them, through the so-called Euler identity ($e^{i\theta} = \cos \theta + i \sin \theta$), with complex numbers (e.g., $3 + 2\sqrt{-1}$). He discovered the imaginary logarithms of negative numbers and showed that each complex number has an infinite number of logarithms.

Euler's textbooks in calculus, *Institutiones calculi differentialis* in 1755 and *Institutiones calculi integralis* in 1768–70, have served as prototypes to the present because they contain formulas of differentiation and numerous methods of indefinite integration, many of which he invented himself, for determining the work done by a force and for solving geometric problems, and he made advances in the theory of linear differential equations, which are useful in solving problems in physics. Thus, he enriched mathematics with substantial new concepts and techniques. He introduced many current notations, such as Σ for the sum; the symbol e for the base of natural logarithms; a, b and c for the sides of a triangle and A, B, and C for the opposite angles; the letter f and parentheses for a function; and i for $\sqrt{-1}$. He also popularized the use of the symbol π (devised by British mathematician William Jones) for the ratio of circumference to diameter in a circle.

After Frederick the Great became less cordial toward him, Euler in 1766 accepted the invitation of Catherine II to return to Russia. Soon after his arrival at St. Petersburg, a cataract formed in his remaining good eye, and he spent the last years of his life in total blindness. Despite this tragedy, his productivity continued undiminished, sustained by an uncommon memory and a remarkable facility in mental computations. His interests were broad, and his *Lettres à une princesse d'Allemagne* in 1768–72 were an admirably clear exposition of the basic principles of mechanics, optics, acoustics, and physical astronomy. Not a classroom teacher, Euler nevertheless had a more pervasive pedagogical influence than any modern mathematician. He had few disciples, but he helped to establish mathematical education in Russia.

Euler devoted considerable attention to developing a more perfect theory of lunar motion, which was particularly troublesome, since it involved the so-called three-body problem—the interactions of Sun, Moon, and Earth. (The problem is still unsolved.) His partial solution, published in 1753, assisted the British Admiralty in calculating lunar tables, of importance then in attempting to determine longitude at sea. One of the feats of his blind years was to perform all the elaborate calculations in his head for his second theory of lunar motion in 1772. Throughout his life Euler was much absorbed by problems dealing with the theory of numbers, which treats of the properties and relationships of integers, or whole numbers (0, ±1, ±2, etc.). In this, his greatest discovery, in 1783, was the law of quadratic reciprocity, which has become an essential part of modern number theory.

In his effort to replace synthetic methods by analytic ones, Euler was succeeded by J.-L. Lagrange. But, where Euler had delighted in special concrete cases, Lagrange sought for abstract generality; and, while Euler incautiously manipulated divergent series, Lagrange attempted to establish infinite processes upon a sound basis. Thus it is that Euler and Lagrange together are regarded as the greatest mathematicians of the 18th century. But Euler has never been excelled either in productivity or in the skillful and imaginative use of algorithmic devices (i.e., computational procedures) for solving problems.

GASPARD MONGE, COUNT DE PÉLUSE
(B. MAY 10, 1746, BEAUNE, FRANCE—
D. JULY 28, 1818, PARIS)

Gaspard Monge, count de Péluse, was a French mathematician who invented descriptive geometry, the study of the mathematical principles of representing three-dimensional objects in a two-dimensional plane. No longer an active discipline in mathematics, the subject is part of mechanical and architectural drawing. He was a prominent figure during the French Revolution, helping to establish the metric system and the École Polytechnique. He was made a count in 1808 by Napoleon I.

Monge was educated at the Oratorian schools at Beaune and at Lyon, where for a time at age 16 he was a physics teacher. He made a large-scale plan of Beaune during a visit in 1762, devising methods of observation and constructing the necessary surveying instruments. Impressed with the plan, a military officer recommended Monge to the commandant of the aristocratic military school of Mézières, where he was accepted as a draftsman.

Gaspard Monge, by Jean Naigeon, 1811; in the Museum of Fine Arts, Beaune, France. Courtesy of the Musée des Beaux-Arts, Beaunc, France.

A further opportunity for Monge to display his skill as a draftsman occurred when he was asked to determine gun emplacements for a proposed fortress. At that time such an operation could be performed only by a long arithmetic process, but Monge devised a geometric method that enabled him to solve the problem so quickly that the commandant at first refused to receive his solution. On later careful examination, Monge's method was classified a military secret. Continuing his research at Mézières, Monge developed his general method of applying geometry to problems of construction. This subject later became known as descriptive geometry and provided an important stimulus to the rediscovery of projective geometry.

Between 1768 and 1783 Monge taught physics and mathematics at Mézières. During this period his main areas of research were in infinitesimal geometry (applications of calculus to geometry) and the theory of partial differential equations. Prompted by the secretary of the French Academy of Sciences, Marie-Jean Condorcet, he wrote a paper discussing the problem of earthworks (composed in 1776 and reworked in 1781) in which he used calculus to determine the curvature of a surface. The paper is of particular importance not for the practical problem it treated but because of its discussion of the theory of surfaces and its introduction of concepts such as the congruence of straight lines and lines of curvature. His work on partial differential equations, characterized by his geometric point of view and in part inspired by the work of Joseph-Louis Lagrange, led him to the development of extremely fruitful new methods. In 1780 Monge was elected an associate of the Academy of Sciences.

Officially leaving Mézières at the end of 1783, Monge became increasingly active in public affairs in Paris. Between 1783 and about 1789 he was an examiner of naval cadets. He served on the committee of weights and measures that established the metric system in 1791. Then from 1792 to 1793 he was minister for the navy and colonies and had occasion to welcome the young artillery officer who became Emperor Napoleon I. In 1795 he participated in the founding of the National Institute of France. Although at times during the French Revolution his position was precarious, Monge continued to be influential. When an appeal was made to scientists to assist in producing materials for national defense, he supervised foundry operations and wrote handbooks on steelmaking and cannon manufacture. In 1794–95 he taught at the short-lived École Normale (later reestablished as the École Normale Supérieure), where he was given permission for the first time to lecture on the principles of descriptive geometry he had developed at Mézières.

Particularly important for mathematics was his substantial role in the founding of the École Polytechnique, which was originally for training engineers and which numbered Lagrange as one of its teachers. Monge was an administrator and an esteemed teacher of descriptive, analytic, and differential geometry. Since no texts were available, his lectures were edited and published for student use. In *Géométrie descriptive* (1799; "Descriptive Geometry"), based on his lectures at the École Normale, he developed his descriptive method for representing a solid in three-dimensional space on a two-dimensional plane by drawing the projections—known as plans, elevations, and traces—of the solid on a sheet of paper. *Feuilles d'analyse appliquée à la géométrie* (1801; "Analysis Applied to Geometry") was an expanded version of his lectures on differential geometry; a later edition incorporated his *Application de l'algèbre à la géométrie* (1805; "Applications of Algebra to Geometry") as *Application de l'analyse à la géométrie* (1807; "Applications of Analysis to Geometry"). Engineering design was revolutionized by his new procedures. Moreover, mathematics education was significantly advanced by his successful texts and popular lectures. Many mathematicians were influenced by his work, notably Jean-Victor Poncelet and Michel Chasles.

Monge was also interested in mechanics and the theory of machines and made contributions to physics and chemistry. In 1796 he became a member of the Commission of Sciences and Arts in Italy and was sent to Italy to choose the paintings and statues that were taken to help finance Napoleon's military campaigns. Many of these works of art went to the Louvre Museum. From 1798 to 1801 he accompanied Napoleon to Egypt, and in Cairo he helped to establish the Institute of Egypt, a cultural organization patterned after the National Institute of France.

With the fall from power of Napoleon in 1814, the Bourbons deprived Monge, a Bonapartist, of all his honours and excluded him in 1816 from the list of members of the reconstituted Institute.

Gilles Personne de Roberval
(b. Aug. 8, 1602, Roberval, France— d. Oct. 27, 1675, Paris)

French mathematician Gilles Personne de Roberval made important advances in the geometry of curves.

In 1632 Roberval became professor of mathematics at the Collège de France, Paris, a position he held until his death. He studied the methods of determination of surface area and volume of solids, developing and improving the method of indivisibles used by the Italian mathematician Bonaventura Cavalieri for computing some of the simpler cases. He discovered a general method of drawing tangents, by treating a curve as the result of the motion of a moving point and by resolving the motion of the point into two simpler components. He also discovered a method for obtaining one curve from another, by means of which planar regions of finite dimensions can be found that are equal in area to the regions between certain curves and their asymptotes (lines that the curves approach but never intersect). To these curves, which were also used to determine areas, the Italian mathematician Evangelista Torricelli gave the name of Robervallian lines.

Roberval indulged in scientific feuds with several of his contemporaries, among them the French philosopher and mathematician René Descartes. He also invented the balance known by his name.

SIMON STEVIN

(B. 1548, BRUGES—D. 1620, THE HAGUE OR LEIDEN, NETH.)

Simon Stevin was a Flemish mathematician who helped standardize the use of decimal fractions and aided in refuting Aristotle's doctrine that heavy bodies fall faster than light ones.

Stevin was a merchant's clerk in Antwerp for a time and eventually rose to become commissioner of public works and quartermaster general of the army under Prince Maurice of Nassau. He engineered a system of sluices to flood certain areas and drive off any enemy, an important defense of Holland. He also invented a 26-passenger carriage with sails for use along the seashore.

In *De Beghinselen der Weeghconst* (1586; "Statics and Hydrostatics") Stevin published the theorem of the triangle of forces. The knowledge of this triangle of forces, equivalent to the parallelogram diagram of forces, gave a new impetus to the study of statics, which had previously been founded on the theory of the lever. He also discovered that the downward pressure of a liquid is independent of the shape of its vessel and depends only on its height and base.

In 1585 Stevin published a small pamphlet, *La Thiend*e ("The Tenth"), in which he presented an elementary and thorough account of decimal fractions and their daily use. Although he did not invent decimal fractions and his notation was rather unwieldy, he established their use in day-to-day mathematics. He declared that the universal introduction of decimal coinage, measures, and weights would be only a question of time. The same year he wrote *La Disme* ("The Decimal") on the same subject.

Stevin published a report in 1586 on his experiment in which two lead spheres, one 10 times as heavy as the other, fell a distance of 30 feet (9.14 metres) in the same time. His report received little attention, though it preceded by three years Galileo's first treatise concerning gravity and by 18 years Galileo's theoretical work on falling bodies.

Modern Geometers

Lars Valerian Ahlfors

(b. April 18, 1907, Helsinki, Fin.—d. Oct. 11, 1996, Pittsfield, Mass., U.S.)

Lars Valerian Ahlfors, a Finnish mathematician, was awarded one of the first two Fields Medals in 1936 for his work with Riemann surfaces. He also won the Wolf Prize in 1981.

Ahlfors received his Ph.D. from the University of Helsinki in 1932. He held an appointment there from 1938 to 1944, then went to the University of Zürich, Switz. He joined the faculty at Harvard University, Cambridge, Mass., U.S., in 1946, remaining there until his retirement. Ahlfors was awarded the Fields Medal at the International Congress of Mathematicians in Oslo, Nor., in 1936. He was cited for methods he had developed to analyze Riemann surfaces of inverse functions in terms of covering surfaces. His principal contributions were in the theory of Riemann surfaces, but his theorems (the Ahlfors finiteness theorem, the Ahlfors five-disk theorem, the Ahlfors principal theorem, etc.) touch on other areas as well, such as the theory of finitely generated Kleinian groups. In 1929 he resolved a conjecture of Arnaud Denjoy on entire functions. Later Ahlfors worked on quasi-conformal mappings and, with Arne Beurling, on conformal invariants.

Ahlfors' publications include *Complex Analysis* (1953); with Leo Sario, Riemann Surfaces (1960); *Lectures on Quasi Conformal Mappings* (1966); and *Conformal Invariants* (1973). His *Collected Papers* was published in 1982.

PAVEL SERGEEVICH ALEKSANDROV

(B. APRIL 25 [MAY 7, NEW STYLE], 1896, BOGORODSK, RUSSIA—D. NOV. 16, 1982, MOSCOW, RUSSIA, U.S.S.R.)

Pavel Sergeevich Aleksandrov was a Russian mathematician who made important contributions to topology.

In 1897 Aleksandrov moved with his family to Smolensk, where his father had accepted a position as a surgeon with the Smolensk State Hospital. His early education was supplied by his mother, who gave him French, German, and music lessons. At grammar school he soon showed an aptitude for mathematics, and on graduation in 1913 he entered Moscow State University.

Aleksandrov had his first major mathematical success in 1915, proving a fundamental theorem in set theory: Every non-denumerable Borel set contains a perfect subset. As often happens in mathematics, the novel ideas used to construct the proof—he invented a way to classify sets according to their complexity—opened new avenues of research. In this case, his ideas provided an important new tool for descriptive set theory.

After graduating in 1917, Aleksandrov moved first to Novgorod-Severskii and then Chernikov to work in the theatre. In 1919, during the Russian Revolution, he was jailed for a short time by White Russians before the Soviet Army recaptured Chernikov. He then returned home to teach at Smolensk State University in 1920 while he prepared for graduate examinations at Moscow State University. On his frequent visits to Moscow he met Pavel Uryson, another graduate student, and began a short but fruitful mathematical collaboration. After passing their graduate exams in 1921, they both became lecturers at Moscow and traveled together throughout western Europe to work with prominent mathematicians each summer from 1922 through 1924, when Uryson drowned in the Atlantic Ocean. On the basis of earlier work by the German mathematician Felix Hausdorff (1869–1942) and others, Aleksandrov and Uryson developed the subject of point-set topology, also known as general topology, which is concerned with the intrinsic properties of various topological spaces. (General topology is sometimes referred to as "rubber-sheet" geometry because it studies those properties of geometric figures or other spaces that are unaltered by twisting and stretching without tearing.) Aleksandrov collaborated with the Dutch mathematician L.E.J. Brouwer during parts of 1925 and 1926 to publish their friend's final papers.

In the late 1920s Aleksandrov developed combinatorial topology, which constructs or characterizes topological spaces through the use of simplexes, a higher dimensional analogy of points, lines, and triangles. He wrote about 300 mathematical books and papers in his long career, including the landmark textbook *Topology* (1935), which was the first and only volume of an intended multivolume collaboration with Swiss mathematician Heinz Hopf.

Aleksandrov was president of the Moscow Mathematical Society (1932–64), vice president of the International Congress of Mathematicians (1958–62), and a full member of the Soviet Academy of Sciences (from 1953). He edited several mathematical journals and received many Soviet awards, including the Stalin Prize (1943) and five Orders of Lenin.

JAMES W. ALEXANDER II

(B. SEPT. 19, 1888, SEA BRIGHT, N.J., U.S.—D. SEPT. 23, 1971, PRINCETON, N.J.)

James W. Alexander II was an American mathematician and a founder of the branch of mathematics originally known as *analysis situs*, now called topology.

The son of John White Alexander, an American painter who created murals for the Library of Congress, James studied mathematics and physics at Princeton University, obtaining a B.S. degree in 1910 and an M.S. degree the following year. For the next few years he traveled and studied in Europe before submitting his doctoral dissertation (1915) to Princeton, where he taught until the United States' entry into World War I in 1917. He was commissioned as a lieutenant in the U.S. Army and served at the Aberdeen Proving Ground in Maryland. Alexander returned to Princeton in 1920, where he remained until 1933 when he joined the newly created Institute for Advanced Studies, a research institution spun off from Princeton. He remained with the institute until his retirement in 1951.Alexander also worked as a civilian consultant for the Army during World War II. Alexander's interest in the relationship of geometric figures that undergo transformation led to his developmental work in topology. His "horned sphere," which is a remarkable deformation of the usual sphere, shows that the topology of three-dimensional space is very different from two-dimensional space. In 1928 Alexander discovered an invariant polynomial, now known as the Alexander polynomial, for distinguishing various knots regardless of how they are stretched or twisted. This was an important first step in providing an algebraic way of distinguishing knots (and therefore three-dimensional manifolds).

Sir Michael Francis Atiyah
(b. April 22, 1929, London, Eng.)

British mathematician Sir Michael Francis Atiyah was awarded the Fields Medal in 1966 primarily for his work in topology. Atiyah received a knighthood in 1983 and the Order of Merit in 1992. He also served as president of the Royal Society (1990–95).

Atiyah's father was Lebanese and his mother Scottish. He attended Victoria College in Egypt and Trinity College, Cambridge (Ph.D., 1955). He held appointments at the Institute for Advanced Study, Princeton, New Jersey, U.S.

Isadore M. Singer, left, and Michael F. Atiyah, middle, receive the Abel Award from Norway's King Harald V during the Abel Award Ceremony in Oslo.

(1955), and at the University of Cambridge (1956–61). In 1961 Atiyah moved to the University of Oxford, where from 1963 to 1969 he held the Savilian Chair of Geometry. He returned to the Institute in 1969 before becoming the Royal Society Research Professor at Oxford in 1972. In 1990 Atiyah became master of Trinity College and director of the Isaac Newton Institute for Mathematical Sciences, both at Cambridge. He retired from the latter position in 1996.

Atiyah was awarded the Fields Medal at the International Congress of Mathematicians in Moscow in 1966 for his work on topology and analysis. He was one of the pioneers, along with the Frenchman Alexandre Grothendieck and the German Friedrich Hirzebruch, in the development of K-theory. This work culminated in 1963, in collaboration with the American Isadore Singer, in the famous

Atiyah-Singer index theorem, which characterizes the number of solutions for an elliptic differential equation. (Atiyah and Singer were jointly recognized for this work with the 2004 Abel Prize.) His early work in topology and algebra was followed by work in a number of different fields, a phenomenon regularly observed in Fields medalists. He contributed, along with others, to the development of the theory of complex manifolds—that is, generalizations of Riemann surfaces to several variables. He also worked on algebraic topology, algebraic varieties, complex analysis, the Yang-Mills equations and gauge theory, and superstring theory in mathematical physics.

Atiyah's publications include *K-theory* (1967); with I.G. Macdonald, *Introduction to Commutative Algebra* (1969); *Elliptic Operators and Compact Groups* (1974); *Geometry of Yang-Mills Fields* (1979); with Nigel Hitchin, *The Geometry and Dynamics of Magnetic Monopoles* (1988); and *The Geometry and Physics of Knots* (1990). His *Collected Works,* in five volumes, appeared in 1988.

EUGENIO BELTRAMI

(B. NOV. 16, 1835, CREMONA, LOMBARDY, AUSTRIAN EMPIRE [NOW IN ITALY]—D. FEB. 18, 1900, ROME, ITALY)

Italian mathematician Eugenio Beltrami was known for his description of non-Euclidean geometry and for his theories of surfaces of constant curvature.

Following his studies at the University of Pavia (1853–56) and later in Milan, Beltrami was invited to join the faculty at the University of Bologna in 1862 as a visiting professor of algebra and analytic geometry. Four years later he was appointed professor of rational mechanics (the application of calculus to the study of the motion of solids and liquids). He also held professorships at universities in Pisa, Rome, and Pavia.

Influenced by the Russian Nikolay Ivanovich Lobachevsky and the Germans Carl Friedrich Gauss and Bernhard Riemann, Beltrami's work on the differential geometry of curves and surfaces removed any doubts about the validity of non-Euclidean geometry, and it was soon taken up by the German Felix Klein, who showed that non-Euclidean geometry was a special case of projective geometry. Beltrami's four-volume work, Opere Matematiche (1902–20), published posthumously, contains his comments on a broad range of physical and mathematical subjects, including thermodynamics, elasticity, magnetism, optics, and electricity. Beltrami was a member of the scientific Accademia dei Lincei, serving as president in 1898. He was elected to the Italian Senate a year before his death.

Enrico Betti

(b. Oct. 21, 1823, Pistoia, Tuscany [Italy]—d. Aug. 11, 1892, Pisa, Kingdom of Italy)

Enrico Betti was a mathematician who wrote a pioneering memoir on topology, the study of surfaces and higher-dimensional spaces, and wrote one of the first rigorous expositions of the theory of equations developed by the noted French mathematician Évariste Galois (1811–32).

Betti studied mathematics and physics at the University of Pisa. After graduating with a degree in mathematics in 1846, he stayed on to work as an assistant until 1849 when he returned home to Pistoia to teach at a secondary school. From 1854 he taught at another secondary school in Florence. In 1857 he obtained a professorship in mathematics at Pisa, where he remained for the rest of his academic life. He fought in two battles for Italian independence and was elected to the new Italian parliament in 1862.

Betti's early work was in the theory of equations and algebra. He extended and furnished proofs for Galois's work, which had previously been stated in part without demonstrations or proofs. (Before Galois could finish his work, he died in a duel at age 21.) The arrival in Pisa of the German mathematician Bernhard Riemann in 1863 decisively affected the course of Betti's research. They became close friends, and Riemann awakened Betti's interest in mathematical physics, in particular potential theory and elasticity, and inspired his memoir on topology. Betti's study of spaces of higher dimensions (greater than three) in the latter work did much to open up the subject, and it led the French mathematician Henri Poincaré to give the name Betti numbers to certain numbers that characterize the connectivity of a manifold (the higher-dimensional analog of a surface).

János Bolyai

(b. Dec. 15, 1802, Kolozsvár, Hungary [now Cluj, Romania]—d. Jan. 27, 1860, Marosvásárhely, Hungary [now Târgu Mureş, Romania])

János Bolyai was a Hungarian mathematician and one of the founders of non-Euclidean geometry—a geometry that differs from Euclidean geometry in its definition of parallel lines. The discovery of a consistent alternative geometry that might correspond to the structure of the universe helped to free mathematicians to study abstract concepts irrespective of any possible connection with the physical world.

By the age of 13, Bolyai had mastered calculus and analytic mechanics under the tutelage of his father, the mathematician Farkas Bolyai. He also became an accomplished violinist at an early age and later was renowned as a superb swordsman. He studied at the Royal Engineering College in Vienna (1818–22) and served in the army engineering corps (1822–33).

The elder Bolyai's preoccupation with proving Euclid's parallel axiom infected his son, and, despite his father's warnings, János persisted in his own search for a solution. In the early 1820s he concluded that a proof was probably impossible and began developing a geometry that did not depend on Euclid's axiom. In 1831 he published "Appendix Scientiam Spatii Absolute Veram Exhibens" ("Appendix Explaining the Absolutely True Science of Space"), a complete and consistent system of non-Euclidean geometry as an appendix to his father's book on geometry, *Tentamen Juventutem Studiosam in Elementa Matheseos Purae Introducendi* (1832; "An Attempt to Introduce Studious Youth to the Elements of Pure Mathematics").

A copy of this work was sent to Carl Friedrich Gauss in Germany, who replied that he had discovered the main results some years before. This was a profound blow to Bolyai, even though Gauss had no claim to priority since he had never published his findings. Bolyai's essay went unnoticed by other mathematicians. In 1848 he discovered that Nikolay Ivanovich Lobachevsky had published an account of virtually the same geometry in 1829.

Although Bolyai continued his mathematical studies, the importance of his work was unrecognized in his lifetime. In addition to work on his non-Euclidean geometry, he developed a geometric concept of complex numbers as ordered pairs of real numbers.

Charles-Julien Brianchon

(b. Dec. 19, 1783, Sèvres, France—
d. April 29, 1864, Versailles)

French mathematician Charles-Julien Brianchon derived a geometrical theorem (now known as Brianchon's theorem) useful in the study of the properties of conic sections (circles, ellipses, parabolas, and hyperbolas) and was innovative in applying the principle of duality to geometry.

In 1804 Brianchon entered the École Polytechnique in Paris, where he became a student of the noted French mathematician Gaspard Monge. While still a student, he published his first paper, "Mémoire sur les surfaces courbes du second degré" (1806; "Memoir on Curved Surfaces of Second Degree"), in which he recognized the projective nature of a theorem of Blaise Pascal, and then proclaimed his own famous theorem: If a hexagon is circumscribed about a conic (all sides made tangent to the conic), then the lines joining the opposite vertices of the hexagon will meet in a single point. The theorem is the dual of Pascal's because its statement and proof can be obtained by systematically substituting the terms *point* with *line* and *collinear* with *concurrent*.

Brianchon graduated first in his class in 1808 and joined Napoleon's armies as a lieutenant in the artillery. Though his courage and ability distinguished him in the field, particularly in the Peninsular War, the rigours of field service affected his health. In 1818 he gained a professorship in the Artillery School of the Royal Guard in Vincennes, where his mathematical work was slowly replaced by other interests.

LUITZEN EGBERTUS JAN BROUWER

(B. FEB. 27, 1881, OVERSCHIE, NETHERLANDS—
D. DEC. 2, 1966, BLARICUM)

Dutch mathematician Luitzen Egbertus Jan Brouwer founded mathematical intuitionism (a doctrine that views the nature of mathematics as mental constructions governed by self-evident laws). In addition, his work completely transformed topology, the study of the most basic properties of geometric surfaces and configurations.

Brouwer studied mathematics at the University of Amsterdam from 1897 to 1904. Even then he was interested in philosophical matters, as evidenced by his *Leven, Kunst, en Mystiek* (1905; "Life, Art, and Mysticism"). In his doctoral thesis, "Over de grondslagen der wiskunde" (1907; "On the Foundations of Mathematics"), Brouwer attacked the logical foundations of mathematics, as represented by the efforts of the German mathematician David Hilbert and the English philosopher Bertrand Russell, and shaped the beginnings of the intuitionist school. The following year, in "Over de onbetrouwbaarheid der logische principes" ("On the Untrustworthiness of the Logical Principles"), he rejected as invalid the use in mathematical proofs of the principle of the excluded middle (or excluded third). According to this principle, every mathematical statement is either true or false; no other possibility is allowed. Brouwer denied that this dichotomy applied to infinite sets.

Brouwer taught at the University of Amsterdam from 1909 to 1951. He did most of his important work in topology between 1909 and 1913. In connection with his studies of the work of Hilbert, he discovered the plane translation theorem, which characterizes topological mappings of the Cartesian plane, and the first of his fixed-point theorems, which later became important in the establishment of some fundamental theorems in branches of mathematics such as differential equations and game theory. In 1911 he established his theorems on the invariance of the dimension of a manifold under continuous invertible transformations. In addition, he merged the methods developed by the German mathematician Georg Cantor with the methods of *analysis situs*, an early stage of topology. In view of his remarkable contributions, many mathematicians consider Brouwer the founder of topology.

In 1918 he published a set theory, the following year a theory of measure, and by 1923 a theory of functions, all developed without using the principle of the excluded middle. He continued his studies until 1954, and, although he did not gain widespread acceptance for his precepts, intuitionism enjoyed a resurgence of interest after World War II, primarily because of contributions by the American mathematician Stephen Cole Kleene.

His *Collected Works*, in two volumes, was published in 1975–76.

MICHEL CHASLES

(B. NOV. 15, 1793, ÉPERNON, FRANCE—D. DEC. 18, 1880, PARIS)

Michel Chasles was a French mathematician who, independently of the Swiss German mathematician Jakob Steiner, elaborated on the theory of modern projective geometry, the study of the properties of a geometric line or other plane figure that remain unchanged when the figure is projected onto a plane from a point not on either the plane or the figure.

Chasles was born near Chartres and entered the École Polytechnique in 1812. He was eventually made professor of geodesy and mechanics there in 1841. His *Aperçu historique sur l'origine et le développement des méthodes en géométrie* (1837; "Historical Survey of the Origin and Development of Geometric Methods") is still a standard historical reference. Its account of projective geometry, including the new theory of duality, which allows geometers to produce new figures from old ones, won the prize of the Academy of Sciences in Brussels in 1829. For its eventual publication Chasles added many valuable historical appendices on Greek and modern geometry.

In 1846 he became professor of higher geometry at the Sorbonne (now one of the Universities of Paris). In that year he solved the problem of determining the gravitational attraction of an ellipsoidal mass to an external point. In 1864 he began publishing in *Comptes rendus*, the journal of the French Academy of Sciences, the solutions to an enormous number of problems based on his "method of characteristics" and his "principle of correspondence." The basis of enumerative geometry is contained in the method of characteristics.

Chasles was a prolific writer and published many of his original memoirs in the *Journal de l'École Polytechnique*. He wrote two textbooks, *Traité de géométrie supérieure* (1852; "Treatise on Higher Geometry") and *Traité des sections coniques* (1865; "Treatise on Conic Sections"). His *Rapport sur le progrès de la géométrie* (1870; "Report on the Progress of Geometry") continues the study in his *Aperçu historique*.

Chasles is also remembered as the victim of a celebrated fraud perpetrated by Denis Vrain-Lucas. He is known to have paid nearly 200,000 francs (approximately $36,000) between 1861 and 1869 for more than 27,000 forged documents—many purported to be from famous men of science, one allegedly a letter from Mary Magdalene to Lazarus, and another a letter from Cleopatra to Julius Caesar—all written in French.

SHIING-SHEN CHERN
(B. OCT. 26, 1911, JIAXING, CHINA—D. DEC. 3, 2004, TIANJIN)

Shiing-shen Chern was a Chinese American mathematician and educator whose research in differential geometry developed ideas that now play a major role in mathematics and in mathematical physics.

Chern graduated from Nankai University in Tianjin, China, in 1930. He received an M.S. degree in 1934 from Tsinghua University in Beijing and a doctor of sciences degree from the University of Hamburg (Germany) in 1936. A year later he returned to Tsinghua as professor of mathematics. Chern was a member of the Institute for Advanced Study at Princeton, New Jersey, from 1943 to 1945. In 1946 he returned to China to become acting director of the Institute of Mathematics at the Academia Sinica in Nanjing.

Chern returned to the United States in 1949 and taught at the University of Chicago, where he collaborated with André Weil, and later at the University of California in Berkeley. In 1961 he became a naturalized U.S. citizen. Chern served as vice president of the American Mathematical Society (1963–64) and was elected to both the National Academy of Sciences and the American Academy of Arts and Sciences. He was awarded the National Medal of Science in 1975 and the Wolf Prize in 1983. He helped found and was the director of the Mathematical Sciences Research Institute in Berkeley (1981–84) and in 1985 played an important role in the establishment of the Nankai Institute of Mathematics in Tianjin, where he held several posts, including director, until his death.

WILLIAM KINGDON CLIFFORD

(B. MAY 4, 1845, EXETER, DEVON, ENG.—D. MARCH 3, 1879, MADEIRA ISLANDS, PORT.)

William Kingdon Clifford was a British philosopher and mathematician who, influenced by the non-Euclidean geometries of Bernhard Riemann and Nikolay Lobachevsky, wrote "On the Space-Theory of Matter" (1876). He presented the idea that matter and energy are simply different types of curvature of space, thus foreshadowing Albert Einstein's general theory of relativity.

Clifford was educated at King's College, London, and at Trinity College, Cambridge, and was elected a fellow of the latter in 1868. In 1871 he was named professor of mathematics at University College, London. Three years later he was elected a fellow of the Royal Society.

Clifford developed the theory of biquaternions (a generalization of the Irish mathematician Sir William Rowan Hamilton's theory of quaternions) and then linked them with more general associative algebras. He used biquaternions to study motion in non-Euclidean spaces and certain closed Euclidean manifolds (surfaces), now known as "spaces of Clifford-Klein." He showed that spaces of constant curvature could have several different topological structures.

Karl Pearson of England further developed Clifford's views on the philosophy of science, which were related to those of Hermann von Helmholtz and Ernst Mach, both of Germany. In philosophy Clifford's name is chiefly associated with two phrases he coined: "mind-stuff" (the simple elements of which consciousness is composed) and "the tribal self." The latter gives the key to his ethical view, which explains conscience and moral law by the development in each individual of a "self" that prescribes conduct conducive to the welfare of the "tribe." He recognized the serious difficulties created for certain features of Immanuel Kant's philosophy by the non-Euclidean geometries of Riemann and Lobachevsky.

Clifford's early death from tuberculosis meant that his works were for the most part published posthumously, and they include *Elements of Dynamic,* 2 vol. (1878, 1887), *Seeing and Thinking* (1879), *Lectures and Essays* (1879), *Math-*

ematical Papers (1882), and *The Common Sense of the Exact Sciences*, completed by Karl Pearson (1885).

Pierre René Deligne
(b. Oct. 3, 1944, Brussels, Belg.)

Belgian mathematician Pierre René Deligne was awarded the Fields Medal at the International Congress of Mathematicians in Helsinki, Fin., in 1978 for his work in algebraic geometry.

Deligne received a bachelor's degree in mathematics (1966) and a doctorate (1968) from the Free University of Brussels. After a year at the National Foundation for Scientific Research, Brussels, he joined the Institute of Advanced Scientific Studies, Bures-sur-Yvette, France, in 1968. In 1984 he became a professor at the Institute for Advanced Study, Princeton, N.J., U.S.

In 1949 the French mathematician André Weil made a series of conjectures concerning zeta functions of curves of abelian varieties. One of these was the equivalent of the Riemann hypothesis for varieties over finite fields. Deligne used a new theory of cohomology called étalestable cohomology, drawing on ideas originally developed by Alexandre Grothendieck some 15 years earlier, and applied them with great success to the Weil conjectures. Deligne's work provided important insights into the relationship between algebraic geometry and algebraic number theory.

He also developed an area of mathematics called weight theory, which has applications in the solution of differential equations. Later, he proved some conjectures by the British topologist Sir William Vallance Douglas Hodge.

Deligne's publications include *Équations différentielles à points singuliers réguliers* (1970; "Differential Equations with Regular Singular Points"); *Groupes de monodromie en géométrie algébrique* (1973; "Monodromy Groups in Algebraic Geometry"); *Modular Functions of One Variable* (1973); with Jean-Francois Boutot et al., *Cohomologie étale* (1977; "Étale Cohomologies"); and, with J. Milne, A. Ogus, and K. Shih, *Hodge Cycles, Motives, and Shimura Varieties* (1982).

Simon Kirwan Donaldson

(b. Aug. 20, 1957, Cambridge, Eng.)

British mathematician Simon Kirwan Donaldson was awarded the Fields Medal in 1986 for his work in topology.

Donaldson attended Pembroke College, Cambridge (B.A., 1979), and Worcester College, Oxford (Ph.D., 1983). From 1983 to 1985 he was a Junior Research Fellow at All Souls College, Oxford, before becoming a fellow and professor at St. Anne's College, Oxford. In 1997 Donaldson joined the faculty of Stanford University in California, U.S., and in 1999 became a professor at Imperial College, London.

Donaldson was awarded the Fields Medal at the International Congress of Mathematicians in Berkeley, California, in 1986, as was the American mathematician Michael Freedman. Their work, taken together, suggested that there are "exotic" four-dimensional spaces—four-dimensional differential manifolds that are topologically equivalent to the standard Euclidean four-dimensional space but that are not equivalent differentiably. This is the only dimension where such exotic spaces exist. It is also an example of a rather common phenomenon in mathematics, wherein problems can be readily solved for all cases beyond a certain number, but for small integers the cases are very complicated and require subtle analysis.

Donaldson's work is rather remarkable in its reversal of the usual direction of ideas from mathematics being applied to solve problems in physics. In particular, Donaldson used the Yang-Mills equations, which are generalizations of James Clerk Maxwell's electro-magnetic equations, to solve problems in pure mathematics. Special solutions to these equations, called instantons, had been applied to physics by earlier mathematicians, but Donaldson used instantons to look at general four-dimensional manifolds. After being awarded the Fields Medal, Donaldson continued his exploitation of ideas from physics with applications to mathematics.

Donaldson's publications include, with Peter Kronheimer, *The Geometry of Four-Manifolds* (1990).

VLADIMIR GERSHONOVICH DRINFELD
(B. FEB. 14, 1954, KHARKOV, UKRAINE, U.S.S.R. [NOW KHARKIV, UKRAINE])

Soviet mathematician Vladimir Gershonovich Drinfeld

was awarded the Fields Medal in 1990 for his work in algebraic geometry and mathematical physics.

Drinfeld attended Moscow State University and the V.A. Steklov Institute of Mathematics, Moscow (Ph.D., 1988). He joined the Institute for Low Temperature Physics and Engineering in Kharkov in 1985. Drinfeld was awarded the Fields Medal at the International Congress of Mathematicians in Kyōto, Japan, in 1990. His principal contributions have been in the theory of automorphic forms, algebraic geometry, and number theory. His interest in the last two led to his working on the Langlands Program, where he solved Langlands' conjecture for a special but important case concerning Galois groups. His work in this area extended earlier explorations by Alexandre Grothendieck, Pierre Deligne, and Robert P. Langlands.

Drinfeld also conducted research in mathematical physics, developing a classification theorem for quantum groups (a subclass of Hopf algebras). He also introduced the ideas of the Poisson-Lie group and Poisson-Lie actions in his work on Yang-Baxter equations, work also related to the quantum groups.

ALEXANDRE GROTHENDIECK

(B. MARCH 28, 1928, BERLIN, GER.—D. NOV. 13, 2014, SAINT-GIRONS, FRANCE)

German French mathematician Alexandre Grothendieck was awarded the Fields Medal in 1966 for his work in algebraic geometry.

After studies at the University of Montpellier (France) and a year at the École Normale Supérieure in Paris, Grothendieck received his doctorate from the University of Nancy (France) in 1953. After appointments at the University of São Paulo in Brazil and the University of Kansas and Harvard University in the United States, he accepted a position at the Institute of Advanced Scientific Studies, Bures-sur-Yvette, France, in 1959. He left in 1970, eventually settling at the University of Montpellier, from which he retired in 1988.

Grothendieck was awarded the Fields Medal at the International Congress of Mathematicians in Moscow in 1966. During the 19th and early 20th centuries there was an enormous growth in the area of algebraic geometry, largely through the tireless efforts of numerous Italian mathematicians. But a more abstract point of view emerged in the mid 20th century, and a great deal of the change is due to the work of Grothendieck, who built on the mathematical work of André Weil, Jean-Pierre Serre, and Oscar Zariski. Using category theory and ideas from topology, he reformulated algebraic geometry so that it applies to commutative rings (such as the integers) and not merely fields (such as the rational numbers) as hitherto. This enabled geometric methods to be applied to problems in number theory and opened up a vast field of research. Among the most notable resulting advances were Gerd Faltings's work on the Mordell conjecture and Andrew Wiles's solution of Fermat's last theorem.

Grothendieck's publications include *Produits tensoriels topologiques et espaces nucléaires* (1955; "Topological Tensor Products and Nuclear Spaces"); with Jean A. Dieudonné, *Éléments de géométrie algébrique* (1960; "Elementary Algebraic Geometry"); and *Espaces vectoriels topologiques* (1973; "Topological Vector Spaces"). A *Festschrift* containing articles in honour of Grothendieck's 60th birthday was published in 1990. Late in his career Grothendieck developed a strong interest in political action; his memoir, *Récoltes et semailles* (1985; "Reaping and Sowing"), is largely concerned with subjects other than mathematics.

DAVID HILBERT

(B. JAN. 23, 1862, KÖNIGSBERG, PRUSSIA [NOW KALININGRAD, RUSSIA]—D. FEB. 14, 1943, GÖTTINGEN, GER.)

David Hilbert was a German mathematician who reduced geometry to a series of axioms and contributed substantially to the establishment of the formalistic foundations of mathematics. His work in 1909 on integral equations led to 20th-century research in functional analysis.

The first steps of Hilbert's career occurred at the University of Königsberg, at which, in 1884, he finished his *Inaugurel-dissertation* (Ph.D.); he remained at Königsberg as a *Privatdozent* (lecturer, or assistant professor) in 1886–92, as an *Extraordinarius* (associate professor) in 1892–93, and as an *Ordinarius* in 1893–95. In 1892 he married Käthe Jerosch, and they had one child, Franz. In 1895 Hilbert accepted a professorship in mathematics at the University of Göttingen, at which he remained for the rest of his life.

The University of Göttingen had a flourishing tradition in mathematics, primarily as the result of the contributions of Carl Friedrich Gauss, Peter Gustav Lejeune Dirichlet, and Bernhard Riemann in the 19th century. During the first three decades of the 20th century this mathematical tradition achieved even greater eminence, largely because of Hilbert. The Mathematical Institute at Göttingen drew students and visitors from all over the world.

Hilbert's intense interest in mathematical physics also contributed to the university's reputation in physics. His colleague and friend, the mathematician Hermann Minkowski, aided in the new application of mathematics to physics until his untimely death in 1909. Three winners of the Nobel Prize for Physics—Max von Laue in 1914, James Franck in 1925, and Werner Heisenberg in 1932— spent significant parts of their careers at the University of Göttingen during Hilbert's lifetime.

In a highly original way, Hilbert extensively modified the mathematics of invariants—the entities that are not altered during such geometric changes as rotation, dilation, and reflection. Hilbert proved the theorem of invariants—that all invariants can be expressed in terms of a finite number. In his *Zahlbericht* ("Commentary on Numbers"), a report on algebraic number theory published in 1897, he consolidated what was known in this subject and pointed the way to the developments that followed. In 1899 he published the *Grundlagen der Geometrie (The Foundations of Geometry*, 1902), which contained his definitive set of axioms for Euclidean geometry and a keen analysis of their significance. This popular book, which appeared in 10 editions, marked a turning point in the axiomatic treatment of geometry.

A substantial part of Hilbert's fame rests on a list of 23 research problems he enunciated in 1900 at the International Mathematical Congress in Paris. In his address, "The Problems of Mathematics," he surveyed nearly all the mathematics of his day and endeavoured to set forth the problems he thought would be significant for mathematicians in the 20th century. Many of the problems have since been solved, and each solution was a noted event. Of those that remain, however, one, in part, requires a solution to the Riemann hypothesis, which is usually considered to be the most important unsolved problem in mathematics.

In 1905 the first award of the Wolfgang Bolyai prize of the Hungarian Academy of Sciences went to Henri Poincaré, but it was accompanied by a special citation for Hilbert.

In 1905 (and again from 1918) Hilbert attempted to lay a firm foundation for mathematics by proving consistency—that is, that finite steps of reasoning in logic could not lead to a contradiction. But in 1931 the Austrian–U.S. mathematician Kurt Gödel showed this goal to be unattainable: propositions may be formulated that are undecidable. Thus, it cannot be known with certainty that mathematical axioms do not lead to contradictions. Nevertheless, the development of logic after Hilbert was different, for he established the formalistic foundations of mathematics.

Hilbert's work in integral equations in about 1909 led directly to 20th-century research in functional analysis (the branch of mathematics in which functions are studied collectively). His work also established the basis for his work on infinite-dimensional space, later called Hilbert space, a concept that is useful in mathematical analysis and quantum mechanics. Making use of his results on integral equations, Hilbert contributed to the development of mathematical physics by his important memoirs on kinetic gas theory and the theory of radiations. In 1909 he proved the conjecture in number theory that for any n, all positive integers are sums of a certain fixed number of nth powers; for example, $5 = 2^2 + 1^2$, in which $n = 2$. In 1910 the second Bolyai award went to Hilbert alone and, appropriately, Poincaré wrote the glowing tribute.

The city of Königsberg in 1930, the year of his retirement from the University of Göttingen, made Hilbert an honorary citizen. For this occasion he prepared an address entitled "Naturerkennen und Logik" ("The Understanding of Nature and Logic"). The last six words of Hilbert's address sum up his enthusiasm for mathematics and the devoted life he spent raising it to a new level: "Wir müssen wissen, wir werden wissen"

("We must know, we shall know"). In 1939 the first Mittag-Leffler prize of the Swedish Academy went jointly to Hilbert and the French mathematician Émile Picard.

The last decade of Hilbert's life was darkened by the tragedy brought to himself and to so many of his students and colleagues by the Nazi regime.

Gaston Maurice Julia

(b. Feb. 3, 1893, Sidi Bel Abbès, Alg.—d. March 19, 1978, Paris, France)

Gaston Maurice Julia was one of the two main inventors of iteration theory and the modern theory of fractals.

Julia emerged as a leading expert in the theory of complex number functions in the years before World War I. In 1915 he exhibited great bravery in the face of a German attack in which he lost his nose and was almost blinded. Awarded the Legion of Honour for his valour, Julia had to wear a black strap across his face for the rest of his life.

Released from service, Julia wrote a memoir on the iteration of polynomial functions (functions whose terms are all multiples of the variable raised to a whole number; e.g., $8x^5 - \sqrt{5}x^2 + 7$) that won the Grand Prix from the French Academy of Sciences in 1918. Together with a similar memoir by French mathematician Pierre Fatou, this created the foundations of the theory. Julia drew attention to a crucial distinction between points that tend to a limiting position as the iteration proceeds and those that never settle down. The former are now said to belong to the Fatou set of the iteration and the latter to the Julia set of the iteration. Julia showed that, except in the simplest cases, the Julia set is infinite, and he described how it is related to the periodic points of the iteration (those that return to themselves after a certain number of iterations). In some cases, this set is the whole plane together with a point at infinity. In other cases, it is a connected curve or is made up entirely of separated points.

After the war, Julia became a professor at the École Polytechnique in Paris, where he ran a major seminar on mathematics and continued to conduct research in geometry and complex function theory. The study of iterative processes in mathematics continued sporadically after Julia's work until the 1970s, when the advent of personal computers enabled mathematicians to produce graphic images of these sets. Stunning colour-coded graphs that showed elaborate structural detail at all scales stimulated a considerable renewal of interest in these objects among both mathematicians and the public.

FELIX KLEIN

(B. APRIL 25, 1849, DÜSSELDORF, PRUSSIA [GERMANY]—D. JUNE 22, 1925, GÖTTINGEN, GER.)

Felix Klein was a German mathematician whose unified view of geometry as the study of the properties of a space that are invariant under a given group of transformations, known as the Erlanger Programm, profoundly influenced mathematical developments.

As a student at the University of Bonn (Ph.D., 1868), Klein worked closely with the physicist and geometer Julius Plücker (1801–68). After Plücker's death, he worked with the geometer Alfred Clebsch (1833–72), who headed the mathematics department at the University of Göttingen. On Clebsch's recommendation, Klein was appointed professor of mathematics at the University of Erlangen (1872–75), where he set forth the views contained in his *Erlanger Programm*. These ideas reflected his close collaboration with the Norwegian mathematician Sophus Lie, whom he met in Berlin in 1869. Before the outbreak of the Franco-German War in July 1870, they were together in Paris developing their early ideas on the role of transformation groups in geometry and on the theory of differential equations.

Klein later taught at the Institute of Technology in Munich (1875–80) and then at the Universities of Leipzig (1880–86) and Göttingen (1886–1913). From 1874 he was the editor of *Mathematische Annalen* ("Annals of Mathematics"), one of the world's leading mathematics journals, and from 1895 he supervised the great *Encyklopädie der mathematischen Wissenschaften mit Einschluss iher Anwendungen* ("Encyclopedia of Pure and Applied Mathematics"). His works on elementary mathematics, including *Elementarmathematik vom höheren Standpunkte aus* (1908; "Elementary Mathematics from an Advanced Standpoint"), reached a wide public. His technical papers were collected in *Gesammelte Math-*

ematische Abhandlungen, 3 vol., (1921–23; "Collected Mathematical Treatises").

Beyond his own work Klein made his greatest impact on mathematics as the principal architect of the modern community of mathematicians at Göttingen, which emerged as one of the world's leading research centres under Klein and David Hilbert (1862–1943) during the period from 1900 to 1914. After Klein's retirement Richard Courant (1888–1972) gradually assumed Klein's role as the organizational leader of this still vibrant community.

NIELS FABIAN HELGE VON KOCH
(B. JAN. 25, 1870, STOCKHOLM, SWED.—
D. MARCH 11, 1924, STOCKHOLM)

Niels Fabian Helge von Koch was a Swedish mathematician famous for his discovery of the von Koch snowflake curve, a continuous curve important in the study of fractal geometry.

Von Koch was a student of Gösta Mittag-Leffler and succeeded him as professor of mathematics at Stockholm University in 1911. His first work was on the theory of determinants of infinite matrices, a topic initiated by the French mathematician Henri Poincaré. This work now forms part of the theory of linear operators, which are fundamental in the study of quantum mechanics. He also worked on the Riemann hypothesis and the prime number theorem.

Von Koch is remembered primarily, however, for a 1906 paper in which he gave a very attractive description of a continuous curve that never has a tangent. Continuous, "nowhere differentiable" functions had been rigorously introduced into mathematics by the German Karl Weierstrass in the 1870s, following suggestions by the German Bernhard Riemann and, even earlier, by the Bohemian Bernhard Bolzano, whose work was not well known. Von Koch's example is perhaps the simplest. Starting with an equilateral triangle, it replaces the middle third of each segment with an equilateral triangle having the deleted portion of the segment as its base (the base is erased). This replacement operation is continued indefinitely, with the result that the limiting curve is continuous but nowhere differentiable. If the new triangles always face outward, the resulting curve will have a striking resemblance to a snowflake, and so the curve is often called von Koch's snowflake.

KODAIRA KUNIHIKO

(B. MARCH 16, 1915, TOKYO, JAPAN—
D. JULY 26, 1997, KŌFU)

Japanese mathematician Kodaira Kunihiko was awarded the Fields Medal in 1954 for his work in algebraic geometry and complex analysis.

Kodaira attended the University of Tokyo (Ph.D., 1949). His dissertation attracted the attention of Hermann Weyl, who invited Kodaira to join him at the Institute for Advanced Study, Princeton, New Jersey,

U.S., where he remained until 1961. After appointments at Harvard University (Cambridge, Massachusetts), Johns Hopkins University (Baltimore, Maryland), and Stanford University (California), he returned to the University of Tokyo in 1967. He retired in 1985.

Kodaira was awarded the Fields Medal at the International Congress of Mathematicians in Amsterdam in 1954. Influenced by Weyl's book on Riemann surfaces, Kodaira conducted research on Riemannian manifolds and Kählerian manifolds. It was in this latter area and in a special subset of these, the Hodge manifolds, that he achieved some of his most important results. In collaboration for many years with the American mathematician D.C. Spencer, he created a theory of the deformation of complex manifolds. Kodaira was principally an algebraic geometer, and his work in this field culminated in his remarkable proof of the Riemann-Roch theorem for functionsofany number ofvariables. Inlater years hedeveloped an interest in the teaching of mathematics and produced, in collaboration with others, a series of mathematics textbooks for elementary and secondary schools.

Kodaira's publications include, with Georges de Rham, *Harmonic Integrals* (1950); with D.C. Spencer, *On Deformations of Complex Analytic Structures* (1957); with James Morrow, *Complex Manifolds* (1971); and *Complex Manifolds and Deformation of Complex Structures* (1986). His *Collected Works* was published in 1975.

Nikolay Ivanovich Lobachevsky

(b. Dec. 1 [Nov. 20, Old Style], 1792, Nizhny Novgorod, Russia—d. Feb. 24 [Feb. 12, Old Style], 1856, Kazan)

Nikolay Ivanovich Lobachevsky was a Russian mathematician and founder of non-Euclidean geometry, which he developed independently of János Bolyai and Carl Gauss. (Lobachevsky's first publication on this subject was in 1829, Bolyai's in 1832. Gauss never published his ideas on non-Euclidean geometry.)

In February 1826 Lobachevsky presented an essay devoted to "the rigorous analysis of the theorem on parallels," in which he may have proposed either a proof of Euclid's fifth postulate (axiom) on parallel lines or an early version of his non-Euclidean geometry. This manuscript, however, was not published, and its content remains unknown. Lobachevsky gave the first public exposition of the ideas of non-Euclidean geometry in his paper "On the principles of geometry," which contained fragments of the 1826 manuscript and was published in 1829–30 in a minor Kazan periodical. In his geometry Lobachevsky abandoned the parallel postulate of Euclid, which states that in the plane formed by a line and a point not on the line it is possible to draw exactly one line through the point that is parallel to the original line. Instead, he based his geometry on the following assumption: In the plane formed by a line and a point not on the line it is possible to draw infinitely many lines through the point that are parallel to the original line. It was later proved that his geometry was self-consistent and, as a result, that the parallel postulate is independent of Euclid's other axioms—hence, not derivable as a theorem from them. This finally resolved an issue that had occupied the minds of mathematicians for over 2,000 years. There can be no parallel theorem, only a parallel postulate. Lobachevsky called his work "imaginary geometry," but as a sympathizer with the empirical spirit of Francis Bacon (1561–1626), he attempted to determine the "true" geometry of space by analyzing astronomical data obtained in the measurement of the parallax of stars. A physical interpretation of Lobachevsky's geometry on a surface of negative curvature was discovered by the Italian mathematician Eugenio Beltrami, but not until 1868.

From 1835 to 1838 Lobachevsky published "Imaginary geometry," "New foundations of geometry with the complete theory of parallels," and "Application of geometry to certain integrals." In 1842 his work was noticed and highly praised by Gauss, at whose instigation Lobachevsky was elected that year as a corresponding member of the Royal Society of Göttingen. Although Lobachevsky was also elected an honorary member of the faculty of Moscow State University, his innovative geometrical ideas provoked misunderstanding and even scorn. The famous Russian mathematician of the time, Mikhail Ostrogradskii, a member of the St. Petersburg Academy, as well as the academician Nicolaus Fuss, spoke pejoratively of Lobachevsky's ideas. Even a literary journal managed to accuse Lobachevsky of "abstruseness." Nevertheless, Lobachevsky continued stubbornly to develop his ideas, albeit in isolation, as he did not maintain close ties with his fellow mathematicians.

In addition to his geometry, Lobachevsky obtained interesting results in algebra and analysis, such as the Lobachevsky–Gräffe method for computing the roots of a polynomial (1834) and the Lobachevsky criterion for convergence of an infinite series (1834–36). His research interests also included the theory of probability, integral calculus, mechanics, astronomy, and meteorology.

The real significance of Lobachevsky's geometry was not fully understood and appreciated until the work of the great German mathematician Bernhard Riemann on the foundations of geometry (1868) and the proof of the consistency of non-Euclidean geometry by his compatriot Felix Klein in 1871. In the late 19th century Kazan State University established a prize and a medal in Lobachevsky's name. Beginning in 1927 the Lobachevsky Prize was awarded by the U.S.S.R. Academy of Sciences (now the Russian Academy of Sciences), but in 1992 the awarding of the medal reverted to Kazan State University.

Benoit Mandelbrot

(b. Nov. 20, 1924, Warsaw, Poland—d. Oct. 14, 2010, Cambridge, Mass., U.S.)

Benoit Mandelbrot is a Polish mathematician universally known as the father of Technology (1947–49). He studied for his doctorate in Paris between 1949 and 1952 and then did research for a year under John von Neumann at the Institute for Advanced Study in Princeton, New Jersey. From 1958 to 1993 he worked for IBM at its Thomas J. Watson Research Center in New York, becoming a research fellow there in 1974. From 1987 he taught at Yale University, becoming the Sterling Professor of Mathematical Sciences in 1999.

As set out in his highly successful book *The Fractal Geometry of Natur*e (1982) and in many articles, Mandelbrot's fractals. Fractals have been employed to describe diverse behaviour in economics, finance, the stock market, astronomy, and computer science.

Mandelbrot was educated at the École Polytechnique (1945–47) in Paris and at the California Institute of work is a stimulating mixture of conjecture and observation, both into mathematical processes and their occurrence in nature and in economics. In 1980 he proposed that a certain set governs the behaviour of some iterative processes in mathematics that are easy to define but have remarkably subtle properties. He produced detailed evidence in support of precise conjectures about this set and helped to generate a substantial and continuing interest in the subject. Many of these conjectures have since been proved by others. The set, now called the Mandelbrot set, has the characteristic properties of a fractal: it is very far from being "smooth," and small regions in the set look like smaller-scale copies of the whole set (a property called self-similarity). Mandelbrot's innovative work with computer graphics stimulated a whole new use of computers in mathematics.

Mandelbrot won a number of awards and honorary degrees. He became a Fellow of the American Academy of Arts and Sciences in 1982 and of the National Academy of Sciences in 1987. He was awarded the Wolf Foundation Prize for Physics in 1993 for his work on fractals, and in 2003 he shared the Japan Prize of the Science and Technology Foundation of Japan for "a substantial contribution to the advance of science and technology."

John Willard Milnor
(b. Feb. 20, 1931, Orange, N.J., U.S.)

American mathematician John Willard Milnor was awarded the Fields Medal in 1962 for his work in differential topology.

Milnor attended Princeton University (A.B., 1951; Ph.D., 1954), in New Jersey. He held an appointment at Princeton from 1954 to 1967 and, after several years at other institutions, joined the faculty at the Institute for Advanced Study, Princeton, in 1970. In 1989 he became director of the Institute for Mathematical Sciences at the State University of New York, Stony Brook.

Milnor was awarded the Fields Medal at the International Congress of Mathematicians in Stockholm in 1962. His work was part of a revival of interest in a geometric approach to topology in the 1950s. Early in the century the field had been highly geometric, but in the

1930s and '40s algebraic approaches dominated research. In particular, Milnor's discovery of multiple differential structures for the seven-dimensional sphere, S7, in 1956 was instrumental in the development of the new field of differential topology.

Additionally, Milnor contributed to algebraic geometry on singular points of complex hypersurfaces, and in 1961 he showed that the *Hauptvermutung* (German: "main conjecture"), a principal conjecture in the theory of manifolds concerning triangulations of *n*-dimensional manifolds, which had been an open question since 1908, is not true for complexes in dimensions greater than 3. Beginning in the 1970s, he worked on complex dynamics.

Milnor was noted as an influential teacher, particularly through his books on the Morse theory and the *h*-cobordism theorem, which are universally regarded as models of mathematical exposition. His publications include *Differential Topology* (1958), *Morse Theory* (1963), *Topology from the Differentiable Viewpoint* (1965), and *Dynamics in One Complex Variable* (1999). His *Collected Papers* were published in five volumes from 1994 to 2010. He won the National Medal of Science in 1966.

HERMANN MINKOWSKI

(B. JUNE 22, 1864, ALEKSOTAS, RUSSIAN EMPIRE [NOW IN KAUNAS, LITHUANIA]—D. JAN. 12, 1909, GÖTTINGEN, GER.)

German mathematician Hermann Minkowski developed the geometrical theory of numbers and made numerous contributions to number theory, mathematical physics, and the theory of relativity. His idea of combining the three dimensions of physical space with that of time into a four-dimensional "Minkowski space"—space-time—laid the mathematical foundations for Albert Einstein's special theory of relativity.

The son of German parents living in Russia, Minkowski returned to Germany with them in 1872 and spent his youth in the royal Prussian city of Königsberg. A gifted prodigy, he began his studies at the University of Königsberg and the University of Berlin at age 15. Three years later he was awarded the "Grand Prix des Sciences Mathématiques" by the French Academy of Sciences for his paper on the representation of numbers as a sum of five squares. During his teenage years in Königsberg he met and befriended another young mathematical prodigy, David Hilbert, with whom he worked closely both at Königsberg and later at the University of Göttingen.

After earning his doctorate in 1885, Minkowski taught mathematics at the Universities of Bonn (1885–94), Königsberg (1894–96), Zürich (1896–1902), and Göttingen (1902–09). Together with Hilbert, he pursued research on the electron theory of the Dutch physicist Hendrik Lorentz and its modification in Einstein's special theory of relativity. In *Raum und Zeit* (1907; "Space and Time") Minkowski gave his famous four-dimensional geometry based on the group of Lorentz transformations of special relativity theory. His major work in number theory was *Geometrie der Zahlen* (1896; "Geometry of Numbers"). His works were collected in David Hilbert (ed.), *Gesammelte Abhandlungen*, 2 vol. (1911; "Collected Papers").

AUGUST FERDINAND MÖBIUS
(B. NOV. 17, 1790, SCHULPFORTA, SAXONY [GERMANY]—D. SEPT. 26, 1868, LEIPZIG)

German mathematician and theoretical astronomer August Ferdinand Möbius is best known for his work in analytic geometry and in topology. In the latter field he is especially remembered as one of the discoverers of the Möbius strip.

Möbius entered the University of Leipzig in 1809 and soon decided to concentrate on mathematics, astronomy, and physics. From 1813 to 1814 he studied theoretical astronomy under Carl Friedrich Gauss at the University of Göttingen. He then studied mathematics at the University of Halle before he obtained a position as a professor of astronomy at Leipzig in 1816. From 1818 to 1821 Möbius supervised the construction of the university's observatory, and in 1848 he was appointed its director.

Möbius's reputation as a theoretical astronomer was established with the publication of his doctoral thesis, *De Computandis Occultationibus Fixarum per Planetas* (1815; "Concerning the Calculation of the Occultations of the Planets"). *Die Hauptsätze der Astronomie* (1836; "The Principles of Astronomy") and *Die Elemente der Mechanik des Himmels* (1843; "The Elements of Celestial Mechanics") are among his other purely astronomical publications.

Möbius's mathematical papers are chiefly geometric; in many of them he developed and applied the methods laid down in his *Der barycentrische Calkul* (1827; "The Calculus of Centres of Gravity"). In this work he introduced homogeneous coordinates (essentially, the extension of coordinates to include a "point at infinity") into analytic geometry and also dealt with geometric transformations, in particular projective transformations that later played an essential part in the systematic development of projective geometry. In the *Lehrbuch der Stati*k (1837; "Textbook on Statics") Möbius gave a geometric treatment of statics, a branch of mechanics concerned with the forces acting on static bodies such as buildings, bridges, and dams.

Möbius was a pioneer in topology. In a memoir of 1865 he discussed the properties of one-sided surfaces, including the Möbius strip produced by giving a narrow strip of material a half-twist before attaching its ends together. Möbius discovered this surface in 1858. The German mathematician Johann Benedict Listing had discovered it a few months earlier, but he did not publish his discovery until 1861. Möbius's *Gesammelten Werke*, 4 vol. ("Collected Works"), appeared in 1885–87.

Mori Shigefumi

(b. Feb. 23, 1951, Nagoya, Japan)

Japanese mathematician Mori Shigefumi was awarded the Fields Medal in 1990 for his work in algebraic geometry.

Mori attended Kyōto University (B.A., 1973; M.A., 1975; Ph.D., 1978) and held an appointment there until 1980, when he went to Nagoya University. In 1990 he joined the faculty of the Research Institute for Mathematical Sciences at Kyōto.

Mori was awarded the Fields Medal at the International Congress of Mathematicians in Kyōto in 1990. In 1979 Mori proved Hartshorne's conjecture, an unsolved problem in algebraic geometry. His most important work focused on the problem of classification of algebraic varieties—solution sets of systems of algebraic equations in some number of variables—in algebraic geometry. The problem of a full classification of algebraic varieties of dimension three was regarded as very difficult, and Mori developed new and powerful techniques to apply to the problem. These problems remain open for higher-dimensional algebraic varieties, although a number of specific results are known.

Mori's publications include, with Herbert Clemens and János Kollár, *Higher Dimensional Complex Geometry* (1988).

DAVID BRYANT MUMFORD

(B. JUNE 11, 1937, WORTH, SUSSEX, ENG.)

British-born mathematician David Bryant Mumford was awarded the Fields Medal in 1974 for his work in algebraic geometry.

Mumford attended Harvard University, Cambridge, Massachusetts, U.S. (B.A., 1957; Ph.D., 1961), staying on to join the faculty upon graduation. He served as vice president (1991–94) and president (1995–98) of the International Mathematical Union. Mumford was awarded the Fields Medal at the International Congress of Mathematicians in Vancouver, British Columbia, Canada, in 1974. As with a number of Fields Medalists, Mumford's prizewinning work was in algebraic geometry. In some of his early work Mumford took up David Hilbert's theory of invariants and applied it to new geometric problems couched in Alexandre Grothendieck's theory of schemes. He continued the efforts of Oscar Zariski in making both algebraic and rigorous the work of the Italian school of algebraic geometers on the subject of algebraic surfaces. He was influential in bringing Grothendieck's ideas to the United States, where they prospered. He also contributed to the development of an algebraic theory of theta functions. In the 1980s Mumford moved to Brown University, Providence, Rhode Island, and began researching the mathematics of computer vision.

Mumford's publications include *Geometric Invariant Theory* (1965) and *Algebraic Geometry* (1976).

Sergey Petrovich Novikov

(b. March 20, 1938, Gorky, Russia, U.S.S.R. [now Nizhny Novgorod, Russia])

Russian mathematician Sergey Petrovich Novikov was awarded the Fields Medal in 1970 for his work in topology.

Novikov graduated from Moscow State University in 1960 and received Ph.D. (1964) and Doctor of Science (1965) degrees from the V.A. Steklov Institute of Mathematics in Moscow. He joined the faculty at Moscow in 1964 and became head of the mathematics department at the L.D. Landau Institute of Theoretical Physics in 1975. In 1983 he became head of the mathematics department at the Steklov Institute. Novikov was awarded the Fields Medal at the International Congress of Mathematicians in Nice, France, in 1970. One of his most impressive contributions in the field of topology was his work on foliations—decompositions of manifolds into smaller ones, called leaves. Leaves can be either open or closed, but at the time Novikov started his work it was not known whether leaves of a closed type existed. Novikov's demonstration of the existence of closed leaves in the case of a three-sphere led to a good deal of additional work in the field. In 1965 he proved the topological invariance of the rational Pontryagin class of differentiable manifolds. He also attacked problems in cohomology and homotopy of Thom spaces of manifolds with striking results. In later years Novikov's work attempted to build bridges between theoretical physics and modern mathematics, particularly in solitons and spectral theory. In addition, he made contributions to algebraic geometry.

Novikov's publications include, with B.A. Dubrovin and A.T. Fomenko, *Sovremennaya geometriya: metody i prilozheniya* (1979; *Modern Geometry: Methods and Applications*) and, with A.T. Fomenko, *Elementi differ-*

entsialnoy geometrii i topologii (1987; *Basic Elements of Differential Geometry and Topology*).

Grigori Perelman
(b. 1966, U.S.S.R.)

Russian mathematician Grigori Perelman was awarded—and declined—the Fields Medal in 2006 for his work on the Poincaré conjecture and Fields medalist William Thurston's geometrization conjecture. In 2003 Perelman had left academia and apparently had abandoned mathematics. He was the first mathematician ever to decline the Fields Medal.

Perelman earned a doctorate from St. Petersburg State University and then spent much of the 1990s in the United States, including at the University of California, Berkeley. He was still listed as a researcher at the Steklov Institute of Mathematics, St. Petersburg University, until Jan. 1, 2006.

In the 1980s, Thurston won a Fields Medal for his efforts to extend the geometric classification of two-dimensional manifolds to three dimensions. Thurston's geometrization conjecture claimed that in three dimensions there are only eight possible geometries, although a three-dimensional manifold may be made up of several regions, each with a different geometry. The conjecture implied that in the particular case of three-dimensional manifolds modeled on the three-dimensional sphere, the Poincaré conjecture is true. In 2000 the Clay Mathematics Institute was formed to stimulate mathematical research by offering $1 million prizes for the solution of important problems in mathematics. The Poincaré conjecture was one of the initial seven Millennium Prize Problems.

In 1982 the American mathematician Richard Hamilton took up the idea of studying how a manifold develops as its curvature is smoothed out, using what is known as a Ricci flow (after the Italian mathematician Gregorio Ricci-Curbastro). Much was achieved, but Hamilton reached an impasse when he could not show that the manifold would not snap into pieces under the flow. Perelman's decisive contribution was to show that the Ricci flow did what was intended and that the impasse reflected the way a three-dimensional manifold is made up of pieces with different geometries. In a series of three difficult papers, published on the Internet in 2002, Perelman announced proofs of the Poincaré conjecture and the geometrization conjecture. By 2006 the consensus among mathematicians was that Perelman had resolved the Poincaré conjecture in the affirmative and likely the geometrization conjecture too. It is generally believed that the techniques he introduced will have a profound influence on other branches of geometry and analysis.

HENRI POINCARÉ

(B. APRIL 29, 1854, NANCY, FRANCE—D. JULY 17, 1912, PARIS)

Henri Poincaré was a French mathematician, and one of the greatest mathematicians and mathematical physicists at the end of 19th century. He made a series of profound innovations in geometry, the theory of differential equations, electromagnetism, topology, and the philosophy of mathematics.

Poincaré grew up in Nancy and studied mathematics from 1873 to 1875 at the École Polytechnique in Paris. He continued his studies at the Mining School in Caen before receiving his doctorate from the École Polytechnique in 1879. While a student, he discovered new types of complex functions that solved a wide variety of differential equations. This major work involved one of the first "mainstream" applications of non-Euclidean geometry, a subject discovered by the Hungarian János Bolyai and the Russian Nikolay Lobachevsky about 1830 but not generally accepted by mathematicians until the 1860s and '70s. Poincaré published a long series of papers on this work in 1880–84 that effectively made his name internationally. The prominent German mathematician Felix Klein, only five years his senior, was already working in the area, and it was widely agreed that Poincaré came out the better from the comparison.

In the 1880s Poincaré also began work on curves defined by a particular type of differential equation, in which he was the first to consider the global nature of the solution curves and their possible singular points (points where the differential equation is not properly defined). He investigated such questions as: Do the solutions spiral into or away from a point? Do they, like the hyperbola, at first approach a point and then swing past and recede from it? Do some solutions form closed loops? If so, do nearby curves spiral toward or away from these closed loops? He showed that the number and types of singular points are determined purely by the topological nature of the surface. In particular, it is only on the torus that the differential equations he was considering have no singular points.

Poincaré intended this preliminary work to lead to the study of the more complicated differential equations that describe the motion of the solar system. In 1885 an added inducement to take the next step presented itself when King Oscar II of Sweden offered a prize for anyone who could establish the stability of the solar system. This would require showing that equations of motion for the planets could be solved and the orbits of the planets shown to be curves that stay in a bounded region of space for all time. Some of the greatest mathematicians since Isaac Newton had attempted to solve this problem, and Poincaré soon realized that he could not make any headway unless he concentrated on a simpler, special case, in which two massive bodies orbit one another in circles around their common centre of gravity while a minute third body orbits them both. The third body is taken to be so small that it does not affect the orbits of the larger ones. Poincaré could establish that the orbit is stable, in the sense that the small body returns infinitely often arbitrarily close to any position it has occupied. This does not mean, however, that it does not also move very far away at times, which would have disastrous consequences for life on Earth. For this and other achievements in his essay, Poincaré was awarded the prize in

1889. But, on writing the essay for publication, Poincaré discovered that another result in it was wrong, and in putting that right he discovered that the motion could be chaotic. He had hoped to show that if the small body could be started off in such a way that it traveled in a closed orbit, then starting it off in almost the same way would result in an orbit that at least stayed close to the original orbit. Instead, he discovered that even small changes in the initial conditions could produce large, unpredictable changes in the resulting orbit. (This phenomenon is now known as pathological sensitivity to initial positions, and it is one of the characteristic signs of a chaotic system.) Poincaré summarized his new mathematical methods in astronomy in *Les Méthodes nouvelles de la mécanique céleste*, 3 vol. (1892, 1893, 1899; "The New Methods of Celestial Mechanics").

Poincaré was led by this work to contemplate mathematical spaces (now called manifolds) in which the position of a point is determined by several coordinates. Very little was known about such manifolds, and, although the German mathematician Bernhard Riemann had hinted at them a generation or more earlier, few had taken the hint. Poincaré took up the task and looked for ways in which such manifolds could be distinguished, thus opening up the whole subject of topology, then known as *analysis situs*. Riemann had shown that in two dimensions surfaces can be distinguished by their genus (the number of holes in the surface), and Enrico Betti in Italy and Walther von Dyck in Germany had extended this work to three dimensions, but much remained to be done. Poincaré singled out the idea of considering closed curves in the manifold that cannot be deformed into one another. For example, any curve on the surface of a sphere can be continuously shrunk to a point, but there are curves on a torus (curves wrapped around a hole, for instance) that cannot. Poincaré asked if a three-dimensional manifold in which every curve can be shrunk to a point is topologically equivalent to a three-dimensional sphere. This problem (now known as the Poincaré conjecture) became one of the most important unsolved problems in algebraic topology. Ironically, the conjecture was first proved for dimensions greater than three: in dimensions five and above by Stephen Smale in the 1960s and in dimension four as a consequence of work by Simon Donaldson and Michael Freedman in the 1980s. Finally, Grigori Perelman proved the conjecture for three dimensions in 2006. All of these achievements were marked with the award of a Fields Medal. Poincaré's *Analysis Situs* (1895) was an early systematic treatment of topology, and he is often called the father of algebraic topology.

Poincaré's main achievement in mathematical physics was his magisterial treatment of the electromagnetic theories of Hermann von Helmholtz, Heinrich Hertz, and Hendrik Lorentz. His interest in this topic—which, he showed, seemed to contradict Newton's laws of mechanics—led him to write a paper in 1905 on the motion of the electron. This paper, and others of his at this time, came close to anticipating Albert Einstein's discovery of the theory of special relativity. But Poincaré never took the decisive step of reformulating traditional concepts of space and time into space-time, which was Einstein's most profound achievement. Attempts were made to obtain a Nobel Prize in physics for Poincaré, but his work was too theoretical and insufficiently experimental for some tastes.

About 1900 Poincaré acquired the habit of writing up accounts of his work in the form of essays and lectures for the general public. Published as *La Science et l'hypothèse* (1903; *Science and Hypothesis*), *La Valeur de la science* (1905; *The Value of Science*), and *Science et méthode* (1908; *Science and Method*), these essays form the core of his reputation as a philosopher of mathematics and science. His most famous claim in this connection is that much of science is a matter of convention. He came to this view on thinking about the nature of space: Was it Euclidean or non-Euclidean? He argued that one could never tell, because one could not logically separate the physics involved from the mathematics, so any choice would be a matter of convention. Poincaré suggested that one would naturally choose to work with the easier hypothesis.

Poincaré's philosophy was thoroughly influenced by psychologism. He was always interested in what the human mind understands, rather than what it can formalize. Thus, although Poincaré recognized that Euclidean and non-Euclidean geometry are equally "true," he argued that our experiences have and will continue to predispose us to formulate physics in terms of Euclidean geometry; Einstein proved him wrong. Poincaré also felt that our understanding of the natural numbers was innate and therefore fundamental, so he was critical of attempts to reduce all of mathematics to symbolic logic (as advocated by Bertrand Russell in England and Louis Couturat in France) and of attempts to reduce mathematics to axiomatic set theory. In these beliefs he turned out to be right, as shown by Kurt Gödel in 1931.

In many ways Poincaré's influence was extraordinary. All the topics discussed above led to the creation of new branches of mathematics that are still highly active today, and he also contributed a large number of more technical results. Yet in other ways his influence was slight. He never attracted a group of students around him, and the younger generation of French mathematicians that came along tended to keep him at a respectful distance. His failure to appreciate Einstein helped to relegate his work in physics to obscurity after the revolutions of special and general relativity. His often imprecise mathematical exposition, masked by a delightful prose style, was alien to the generation in the 1930s who modernized French mathematics under the collective pseudonym of Nicolas Bourbaki, and they proved to be a powerful force. His philosophy of mathematics lacked the technical aspect and profundity of developments inspired by the German mathematician David Hilbert's work. However, its diversity and fecundity has begun to prove attractive again in a world that sets more store by applicable mathematics and less by systematic theory.

Most of Poincaré's original papers are published in the 11 volumes of his *Oeuvres de Henri Poincaré* (1916–54). In 1992 the Archives–Centre d'Études et de Recherche Henri-Poincaré founded at the University of Nancy 2 began to edit Poincaré's scientific correspondence, signaling a resurgence of interest in him.

JEAN-VICTOR PONCELET

(B. JULY 1, 1788, METZ, FRANCE—D. DEC. 22, 1867, PARIS)

French mathematician and engineer Jean-Victor Poncelet was one of the founders of modern projective geometry. His principle of continuity, a concept designed to add generality to synthetic geometry (limited to geometric arguments), led to the introduction of imaginary points and the development of algebraic geometry.

As a lieutenant of engineers in 1812, he took part in Napoleon's Russian campaign, in which he was abandoned as dead at Krasnoy and imprisoned at Saratov; he returned to France in 1814. During his imprisonment Poncelet studied projective geometry and wrote *Applications d'analyse et de géométrie*, 2 vol. (1862–64; "Applications of Analysis and Geometry"). This work was originally planned as an introduction to his celebrated *Traité des propriétés projectives des figures* (1822; "Treatise on the Projective Properties of Figures"), for which Poncelet is regarded as one of the greatest projective geometers. His development of the pole and polar lines associated with conic sections led to the principle of duality (exchanging "dual" elements, such as points and lines, along with their corresponding statements, in a true theorem produces a true "dual statement") and a dispute over priority with the German mathematician Julius Plücker for its discovery. His principle of continuity, a concept designed to add generality to synthetic geometry (limited to geometric arguments), led to the introduction of imaginary points and the development of algebraic geometry.

From 1815 to 1825 Poncelet was occupied with military engineering at Metz, and from 1825 to 1835 he was a professor of mechanics at the École d'Application there. He applied mathematics to the improvement of turbines and waterwheels. Although the first inward-flow turbine was not built until 1838, he proposed such a turbine in 1826. In Paris from 1838 to 1848 he was a professor at the Faculty of Sciences, and from 1848 to 1850 he was commandant of the École Polytechnique, with the rank of general.

BERNHARD RIEMANN

(B. SEPT. 17, 1826, BRESELENZ, HANOVER [GERMANY]—
D. JULY 20, 1866, SELASCA, ITALY)

Bernhard Riemann was a German mathematician whose profound and novel approaches to the study of geometry laid the mathematical foundation for Albert Einstein's theory of relativity. He also made important contributions to the theory of functions, complex analysis, and number theory.

Riemann was born into a poor Lutheran pastor's family, and all his life he was a shy and introverted person. He was fortunate to have a schoolteacher who recognized his rare mathematical ability and lent him advanced books to read, including Adrien-Marie Legendre's *Number Theory* (1830). Riemann read the book in a week and then claimed to know it by heart. He went on to study mathematics at the University of Göttingen in 1846–47 and 1849–51 and at the University of Berlin (now the Humboldt University of Berlin) in 1847–49. He then gradually worked his way up the academic profession, through a succession of poorly paid jobs, until he became a full professor in 1859 and gained, for the first time in his life, a measure of financial security. However, in 1862, shortly after his marriage to Elise Koch, Riemann fell

Bernhard Riemann, lithograph after a portrait, artist unknown, 1863. Archiv für Kunst und Geschichte, Berlin.

seriously ill with tuberculosis. Repeated trips to Italy failed to stem the progress of the disease, and he died in Italy in 1866.

Riemann's visits to Italy were important for the growth of modern mathematics there; Enrico Betti in particular took up the study of Riemannian ideas. Ill health prevented Riemann from publishing all his work, and some of his best was published only posthumously—e.g., the first edition of Riemann's *Gesammelte mathematische Werke* (1876; "Collected Mathematical Works"), edited by Richard Dedekind and Heinrich Weber.

Riemann's influence was initially less than it might have been. Göttingen was a small university, Riemann was a poor lecturer, and, to make matters worse, several of his best students died young. His few papers are also difficult to read, but his work won the respect of some of the best mathematicians in Germany, including his friend Dedekind and his rival in Berlin, Karl Weierstrass. Other mathematicians were gradually drawn to his papers by their intellectual depth, and in this way he set an agenda for conceptual thinking over ingenious calculation. This emphasis was taken up by Felix Klein and David Hilbert, who later established Göttingen as a world centre for mathematics research, with Carl Gauss and Riemann as its iconic figures.

In his doctoral thesis (1851), Riemann introduced a way of generalizing the study of polynomial equations in two real variables to the case of two complex variables.

In the real case a polynomial equation defines a curve in the plane. Because a complex variable z can be thought of as a pair of real variables $x + iy$ (where $i = \sqrt{-1}$), an equation involving two complex variables defines a real surface—now known as a Riemann surface—spread out over the plane. In 1851 and in his more widely available paper of 1857, Riemann showed how such surfaces can be classified by a number, later called the genus, that is determined by the maximal number of closed curves that can be drawn on the surface without splitting it into separate pieces. This is one of the first significant uses of topology in mathematics.

In 1854 Riemann presented his ideas on geometry for the official postdoctoral qualification at Göttingen; the elderly Gauss was an examiner and was greatly impressed. Riemann argued that the fundamental ingredients for geometry are a space of points (called today a manifold) and a way of measuring distances along curves in the space. He argued that the space need not be ordinary Euclidean space and that it could have any dimension (he even contemplated spaces of infinite dimension). Nor is it necessary that the surface be drawn in its entirety in three-dimensional space. A few years later this inspired the Italian mathematician Eugenio Beltrami to produce just such a description of non-Euclidean geometry, the first physically plausible alternative to Euclidean geometry. Riemann's ideas went further and turned out to provide the mathematical foundation for the four-dimensional geometry of space-time in Einstein's theory of general relativity. It seems that Riemann was led to these ideas partly by his dislike of the concept of action at a distance in contemporary physics and by his wish to endow space with the ability to transmit forces such as electromagnetism and gravitation.

In 1859 Riemann also introduced complex function theory into number theory. He took the zeta function, which had been studied by many previous mathematicians because of its connection to the prime numbers, and showed how to think of it as a complex function. The Riemann zeta function then takes the value zero at the negative even integers (the so-called trivial zeros) and also at points on a certain line (called the critical line). Standard methods in complex function theory, due to Augustin-Louis Cauchy in France and Riemann himself, would give much information about the distribution of prime numbers if it could be shown that all the nontrivial zeros lie on this line—a conjecture known as the Riemann hypothesis. All nontrivial zeros discovered thus far have been on the critical line. In fact, infinitely many zeros have been discovered to lie on this line. Such partial results have been enough to show that the number of prime numbers less than any number x is well approximated by $x/\ln x$. The Riemann hypothesis was one of the 23 problems that Hilbert challenged mathematicians to solve in his famous 1900 address, "The Problems of Mathematics." Over the years a growing body of mathematical ideas have built upon the assumption that the Riemann hypothesis is true; its proof, or disproof, would have far-reaching consequences and confer instant renown.

Riemann took a novel view of what it means for mathematical objects to exist. He sought general existence proofs, rather than "constructive proofs" that actually produce the objects. He believed that this approach led to conceptual clarity and prevented the mathematician from getting lost in the details, but even some experts disagreed with such nonconstructive proofs. Riemann also studied how functions compare with their trigonometric or Fourier series representation, which led him to refine ideas about discontinuous functions. He showed how complex function theory illuminates the study of minimal surfaces (surfaces of least area that span a given boundary). He was one of the first to study differential equations involving complex variables, and his work led to a profound connection with group theory. He introduced new general methods in the study of partial differential equations and applied them to produce the first major study of shock waves.

Jean-Pierre Serre
(b. Sept. 15, 1926, Bages, France)

French mathematician Jean-Pierre Serre was awarded the Fields Medal in 1954 for his work in algebraic topology. In 2003 he was awarded the first Abel Prize by the Norwegian Academy of Science and Letters.

Serre attended the École Normale Supérieure (1945–48) and the Sorbonne (Ph.D.; 1951), both now part of the Universities of Paris. Between 1948 and 1954 he was at the National Centre for Scientific Research in Paris, and after two years at the University of Nancy he returned to Paris for a position at the Collège de France. He retired in 1994. Between 1983 and 1986 Serre served as vice president of the International Mathematical Union.

Serre was awarded the Fields Medal at the International Congress of Mathematicians in Amsterdam in 1954. Serre's mathematical contributions leading up to the Fields Medal were largely in the field of algebraic topology, but his later work ranged widely—in algebraic geometry, group theory, and especially number theory. By seeing unifying ideas, he helped to unite disparate branches of mathematics. One of the more recent phenomena in which he was a principal contributor was the applications of algebraic geometry to number theory—applications now falling into a separate subclass called arithmetic geometry. He was one of the second generation of members of Nicolas Bourbaki (publishing pseudonym for a group of mathematicians) and a source of inspiration for fellow medalists Alexandre Grothendieck and Pierre Deligne.

An elegant writer of mathematics, *Serre published Groupes algébriques et corps de classes* (1959; *Algebraic Groups and Class Fields*); *Corps locaux* (1962; *Local Fields*); *Lie Algebras and Lie Groups* (1965); *Abelian l-adic Representations and Elliptic Curves* (1968); *Cours d'arithmétique* (1970; *A Course in Arithmetic*); *Cohomologie Galoisienne* (1964; *Galois Cohomology*); *Représentations linéaires des groupes finis* (1967; *Linear Representations of Finite Groups*); *Algèbre locale, multiplicités* (1965; "Local Algebra: Multiplicities"); *Arbres, amalgames, SL^2* (1977; *Trees*); and, with Uwe Jannsen and Steven L. Kleiman, *Motives* (1994). His collected works were published in 1986. A Leroy P. Steele Prize in 1995 was awarded to Serre on the basis of *A Course in Arithmetic*.

Wacław Sierpiński

(b. March 14, 1882, Warsaw, Russian Empire [now in Poland]—d. Oct. 21, 1969, Warsaw)

Wacław Sierpiński was a leading figure in point-set topology and one of the founding fathers of the Polish school of mathematics, which flourished between World Wars I and II.

Sierpiński graduated from Warsaw University in 1904, and in 1908 he became the first person anywhere to lecture on set theory. During World War I it became clear that an independent Polish state might emerge, and Sierpiński, with Zygmunt Janiszewski and Stefan Mazurkiewicz, planned the future shape of the Polish mathematical community: it would be centred in Warsaw and Lvov, and, because resources for books and journals would be scarce, research would be concentrated in set theory, point-set topology, the theory of real functions, and logic. Janiszewski died in 1920, but Sierpiński and Mazurkiewicz successfully saw the plan through. At the time it seemed a narrow and even risky choice of topics, but it proved highly fruitful, and a stream of fundamental work in these areas came out of Poland until the intellectual life of the country was destroyed by the Nazis and the invading Soviet forces.

Sierpiński's own work in set theory and topology was extensive, amounting to over 600 research papers, and toward the end of his life he added a further 100 papers on number theory. He expended much effort on giving a topological characterization of the continuum (the set of real numbers) and in this way discovered many examples of topological spaces with unexpected properties, of which the Sierpiński gasket is the most famous. The Sierpiński gasket is defined as follows: Take a solid equilateral triangle, divide it into four congruent equilateral triangles, and remove the middle triangle; then do the same with each of the three remaining triangles; and so on. The resulting fractal is self-similar (small parts of it are scale copies of the whole thing). Also, it has an area of zero, a fractional dimension (between a one-dimensional line and a two-dimensional plane figure), and a boundary of infinite length. A similar construction starting with a square produces the Sierpiński carpet, which is also self-similar. Good approximations of these and other fractals have been used to produce compact multiband radio antennas.

Stephen Smale
(b. July 15, 1930, Flint, Mich., U.S.)

American mathematician Stephen Smale was awarded the Fields Medal in 1966 for his work on topology in higher dimensions.

Smale grew up in a rural area near Flint. From 1948 to 1956 he attended the University of Michigan, obtaining B.S., M.S., and Ph.D. degrees in mathematics. As an instructor at the University of Chicago from 1956 to 1958, Smale achieved notoriety by proving that there exists an eversion of the sphere (meaning, in a precise theoretical sense, that it is possible to turn a sphere inside out).

In 1960 Smale obtained his two most famous mathematical results. First he constructed a function, now known as the horseshoe, that serves as a paradigm for chaos. Next Smale proved the generalized Poincaré conjecture for all dimensions greater than or equal to five. (The classical conjecture states that a simply connected closed three-dimensional manifold is a three-dimensional sphere, a set of points in four-dimensional space at the same distance from the origin.) The two-dimensional version of this theorem (the two-dimensional sphere is the surface of a common sphere in three-dimensional space) was established in the 19th century, and the three-dimensional version was established at the start of the 21st century. Smale's work was remarkable in that he bypassed dimensions three and four to resolve the problem for all higher dimensions. In 1961 he followed up with the *h*-cobordism theorem, which became the fundamental tool for classifying different manifolds in higher-dimensional topology.

In 1965 Smale took a six-month hiatus from mathematical research to join radical activist Jerry Rubin in establishing the first campaign of nonviolent civil disobedience directed at ending U.S. involvement in the Vietnam War. Smale's mathematical and political lives collided the following year at the International Congress of Mathematicians in Moscow, where he received the Fields Medal. There Smale held a controversial press conference in which he criticized the actions of both the U.S. and Soviet governments.

Smale's mathematical work is notable for both its breadth and depth, reaching the areas of topology, dynamical systems, economics, nonlinear analysis, mechanics, and computation. In 1994 Smale retired from the University of California at Berkeley and then joined the faculty of the City University of Hong Kong.

Smale's publications include *Differential Equations, Dynamical Systems, and Linear Algebra* (1974; with Morris W. Hirsch), *The Mathematics of Time: Essays on Dynamical Systems, Economic Processes, and Related Topics* (1980), and *The Collected Papers of Stephen Smale* (2000).

Karl Georg Christian von Staudt

(b. Jan. 24, 1798, Imperial Free City of Rothenburg [now Rothenburg ob der Tauber, Ger.]—d. June 1, 1867, Erlangen, Bavaria)

German mathematician Karl Georg Christian von Staudt developed the first purely synthetic theory of imaginary points, lines, and planes in projective geometry. Later geometers, especially Felix Klein (1849–1925), Moritz Pasch (1843–1930), and David Hilbert (1862–1943), exploited these possibilities for bridging the gap between synthetic and analytic methods in geometry.

Staudt studied mathematics and astronomy under Carl Gauss from 1818 to 1822 at the University of Göttingen. Under Gauss's supervision he published research on ephemerides and the orbits of asteroids. After taking his doctorate at the University of Erlangen in 1822, Staudt taught mathematics at the Gymnasium in Nuremberg from 1827 to 1835. From 1835 until his death he was professor of mathematics at the University of Erlangen.

Although Staudt's principal fame was due to his contributions to the geometry of position (now known as projective geometry), he also worked on the arithmetical properties of Bernoulli numbers and on the Von Staudt–Clausen theorem. His main works were *Geometrie der Lage* (1847; "The Geometry of Position") and *Beiträge zur Geometrie der Lage* (1856–60; "Contributions to the Geometry of Position").

JAKOB STEINER
(B. MARCH 18, 1796, UTZENSTORF, SWITZ.—
D. APRIL 1, 1863, BERN)

Swiss mathematician Jakob Steiner was one of the founders of modern synthetic and projective geometry.

As the son of a small farmer, Steiner had no early schooling and did not learn to write until he was 14. Against the wishes of his parents, at 18 he entered the Pestalozzi School at Yverdon, Switzerland, where his extraordinary geometric intuition was discovered. Later he went to the University of Heidelberg and the University of Berlin to study, supporting himself precariously as a tutor. By 1824 he had studied the geometric transformations that led him to the theory of inversive geometry, but he did not publish this work. The founding in 1826 of the first regular publication devoted to mathematics, *Crelle's Journal*, gave Steiner an opportunity to publish some of his other original geometric discoveries. In 1832 he received an honorary doctorate from the University of Königsberg, and two years later he occupied the chair of geometry established for him at Berlin, a post he held until his death.

During his lifetime some considered Steiner the greatest geometer since Apollonius of Perga (c. 262–190 BCE), and his works on synthetic geometry were considered authoritative. He had an extreme dislike for the use of algebra and analysis, and he often expressed the opinion that calculation hampered thinking, whereas pure geometry stimulated creative thought. By the end of the century, however, it was generally recognized that Karl von Staudt (1798–1867), who worked in relative isolation at the University of Erlangen, had made far deeper contributions to a systematic theory of pure geometry. Nevertheless, Steiner contributed many basic concepts and results in projective geometry. For example, during a trip to Rome in 1844 he discovered a transformation of the real projective plane (the set of lines through the origin in ordinary three-dimensional space) that maps each line of the projective plane to one point on the Steiner surface (also known as the Roman surface). Steiner never published these and other findings concerning the surface. A colleague, Karl Weierstrass, first published a paper on the surface and Steiner's results in 1863, the year of Steiner's death. Steiner's other work was primarily on the properties of algebraic curves and surfaces and on the solution of isoperimetric problems. His collected writings were published posthumously as *Gesammelte Werke*, 2 vol. (1881–82; "Collected Works").

René Frédéric Thom

(b. Sept. 2, 1923, Montbéliard, France—d. Oct. 25, 2002, Bures-sur-Yvette)

René Frédéric Thom was a French mathematician who was awarded the Fields Medal in 1958 for his work in topology.

Thom graduated from the École Normale Supérieure (now part of the Universities of Paris) in 1946, spent four years at the nearby National Centre for Scientific Research, and in 1951 was awarded a doctorate by the University of Paris. He held appointments at the University of Grenoble (1953–54) and the University of Strasbourg (1954–63). In 1964 he became a professor at the Institute of Advanced Scientific Studies, Bures-sur-Yvette.

Thom was awarded the Fields Medal at the International Congress of Mathematicians in Edinburgh in 1958 for his numerous important contributions in topology, particularly the introduction of the concept of cobordism. Cobordism is a tool for classifying differentiable manifolds. Two manifolds of dimension n are cobordant if there exists a manifold-with boundary of dimension $n + 1$, whose boundary is their disjoint union. He is best known, however, for catastrophe theory, an attempt to model abrupt behavioral changes—such as the transition from liquid to gas or, in human events, from peace to war—with functions on surfaces that have folds and cusps. The mathematical insight was valuable, but the subject became controversial when some of Thom's friends and colleagues made rather extravagant claims on the applicability of catastrophe theory. In addition, it was realized that many of the associated ideas, under different terminology, had already been employed by applied mathematicians.

Thom's publications include *Stabilité structurelle et morphogénèse* (1972; Structural Stability and Morphogenesis) and *Théorie des catastrophes et biologie* (1979; "Catastrophe Theory in Biology").

William Paul Thurston
(b. Oct. 30, 1946, Washington, D.C., U.S.)

American mathematician William Paul Thurston was awarded the Fields Medal in 1983 for his work in topology.

Thurston was educated at New College, Sarasota, Florida (B.A., 1967), and the University of California, Berkeley (Ph.D., 1972). After a year at the Institute for Advanced Study, Princeton, New Jersey, he joined the faculty of the Massachusetts Institute of Technology (1973–74) and then moved to Princeton University, where he remained until 1991. In 1992 he became director of the Mathematical Sciences Research Institute at Berkeley. In 1996 he moved to the University of California, Davis.

Thurston was awarded the Fields Medal at the International Congress of Mathematicians in Warsaw in 1983 for his work in the topology of two and three dimensions. He extended geometric ideas from the theory of two-dimensional manifolds to the study of three-dimensional manifolds. His geometrization conjecture says that every three-dimensional manifold is locally isometric to just one of a family of eight distinct types. Special cases were then proved, but only in 2006 was the first generally convincing proof published of the Poincaré conjecture in three dimensions, which was a major unresolved part of Thurston's geometrization conjecture. Grigori Perelman was awarded a Fields Medal in 2006 for this achievement, which built on earlier work of Richard Hamilton, and for his proof of the full geometrization conjecture. Thurston also took up ideas about the discrete isometry groups of hyperbolic three-space, first investigated by Henri Poincaré and later studied by Lars Ahlfors. Deformations of these groups were studied by Thurston, and further advances in quasi-conformal maps resulted.

Thurston was an enthusiast for an unusual style of mathematical writing that was strong on intuition and short on proofs. His publications included *The Geometry and Topology of 3-Manifolds* (1979) and *Three-Dimensional Geometry and Topology* (1997).

OSWALD VEBLEN
(B. JUNE 24, 1880, DECORAH, IOWA, U.S.—D. AUG. 10, 1960, BROOKLIN, MAINE)

American mathematician Oswald Veblen made important contributions to differential geometry and the early development of topology. Many of his contributions found application in atomic physics and the theory of relativity.

Veblen graduated from the University of Iowa in 1898. He spent a year at Harvard University before moving to the University of Chicago (Ph.D., 1903). He taught mathematics at Princeton University (1905–32) and was appointed a professor at the Institute for Advanced Study, Princeton, New Jersey, when it opened in 1932. Veblen played a key role in the formation and research direction of the school of mathematics at the institute. He became professor emeritus in 1950.

From the beginning of his research career, Veblen was interested in the foundations of mathematics. His thesis concerned the axiomatization of Euclidean geometry and had implications in the study of mathematical logic. This began his study of axiom systems in projective geometry, which culminated in the highly acclaimed *Projective Geometry*, 2 vol. (1910–18), in collaboration with John Wesley Young.

Veblen's *Analysis Situs* (1922) was the first book to cover the basic ideas of topology systematically. It was his most influential work and for many years the best available topology text. Veblen also laid the foundations for topological research at Princeton.

Soon after the discovery of general relativity, Veblen turned to differential geometry and took a leading part in the development of generalized affine and projective geometry. His work *The Invariants of Quadratic Differential Forms* (1927) is distinguished by precise and systematic treatment of the basic properties of Riemannian geometry. In collaboration with his brilliant student John Henry Whitehead, Veblen extended the knowledge of the Riemann metric for more general cases in *The Foundations of Differential Geometry* (1932).

Veblen's belief that "the foundations of geometry must be studied both as a branch of physics and as a branch of mathematics" quite naturally led him to the study of relativity and the search for a geometric structure to form a field theory unifying gravitation and electromagnetism. With respect to the Kaluza-Klein field theory, which involved field equations in five-dimensional space, he provided the first physical interpretation of the fifth coordinate. By regarding the coordinate as a gauge variable, he was able to interpret the theory as one involving four-dimensional space-time. In connection with this contribution, Veblen provided a new treatment of spinors (expressions used to represent electron spin) that he summarized in *Projektive Relativitätstheorie* (1933; "Projective Relativity Theory").

Veblen was notable in his efforts to aid German mathematicians displaced by the Nazi regime. These activities, combined with his tremendous influence in encouraging and developing young mathematicians, represent a contribution equal to that of his mathematical innovations.

VLADIMIR VOEVODSKY

(B. JUNE 4, 1966, MOSCOW, RUSSIA, U.S.S.R.)

Russian mathematician Vladimir Voevodsky won the Fields Medal in 2002 for having made one of the most outstanding advances in algebraic geometry in several decades.

Voevodsky attended Moscow State University (1983–89) before earning a Ph.D. from Harvard University in 1992. He then held visiting positions at Harvard (1993–96) and at Northwestern University, Evanston, Illinois (1996–98), before becoming a permanent professor in 1998 at the Institute for Advanced Study, Princeton, N.J.

Voevodsky was awarded the Fields Medal at the International Congress of Mathematicians in Beijing in 2002. In an area of mathematics noted for its abstraction, his work is particularly praised for the ease and flexibility with which he has deployed it in solving quite concrete mathematical problems. Voevodsky built on the work of one of the most influential mathematicians of the 20th century, the 1966 Fields Medalist Alexandre Grothendieck. Grothendieck proposed a novel mathematical structure ("motives") that would enable algebraic geometry to adopt and adapt methods used with great success in algebraic topology. Algebraic topology applies algebraic techniques to the study of topology, which concerns those essential aspects of objects (such as the number of holes) that are not changed by any deformation (stretching, shrinking, and twisting with no tearing). In contrast, algebraic geometry applies algebraic techniques to the study of rigid shapes. It has proved much harder in this discipline to identify essential features in a usable way. In a major advancement of Grothendieck's program for unifying these vast regions of mathematics, Voevodsky proposed a new way of working with motives, using new cohomology theories. His work has important ramifications for many different topics in number theory and algebraic geometry.

André Weil

(b. May 6, 1906, Paris, France—
d. Aug. 6, 1998, Princeton, N.J., U.S.)

André Weil was a French mathematician who was one of the most influential figures in mathematics during the 20th century, particularly in number theory and algebraic geometry.

André was the brother of the philosopher and mystic Simone Weil. He studied at the École Normale Supérieure (now part of the Universities of Paris) and at the Universities of Rome and Göttingen, receiving his doctorate from the University of Paris in 1928. His teaching career was even more international. He was professor of mathematics at the Aligarh Muslim University, India (1930–32), and thereafter taught at the University of Strasbourg, France (1933–40), the University of São Paulo, Brazil (1945–47), and the University of Chicago (1947–58). He joined the Institute for Advanced Study, Princeton, N.J., U.S., in 1958, becoming professor emeritus in 1976. He was also a gifted linguist who read Sanskrit and many other languages, and he was a sympathetic expert on Indian religious writings.

Beginning in the mid 1930s, as one of the founding members of a group of French mathematicians writing under the collective pseudonym Nicolas Bourbaki, Weil worked and inspired others in the effort to achieve David Hilbert's program of unifying all of mathematics upon a rigorous axiomatic basis and directed to the solution of significant problems. Weil and Jean Dieudonné were chiefly responsible for Bourbaki's interest in the history of mathematics, and Weil wrote on it extensively toward the end of his career.

Weil made fundamental contributions to algebraic geometry—at that time a subject mostly contributed to by members of the "Italian school" but being reformulated along algebraic lines by Bartel van der Waerden and Oscar Zariski—and algebraic topology. Weil believed that many fundamental theorems in number theory and algebra had analogous formulations in algebraic geometry and topology. Collectively known as the Weil conjectures, they became the basis for both these disciplines. In particular, Weil began the proof of a variant of the Riemann hypothesis for algebraic curves while interned in Rouen, France, in 1940 for his deliberate failure, as a pacifist, to report for duty in the French army. This internment followed his incarceration and later expulsion from Finland, where he was suspected of being a spy. In order to avoid a five-year sentence in a French jail, Weil volunteered to return to the army. In 1941, after reuniting with his wife, Eveline, Weil fled with her to the United States.

The Weil conjectures generated many new ideas in algebraic topology. Their importance can be gauged by the fact that the Belgian mathematician Pierre Deligne was awarded a Fields Medal in 1978 in part for having proved one of the conjectures. The Weil conjectures have recently had ramifications in cryptology, computer modeling, data transmission, and other fields.

Weil's published works include *Foundations of Algebraic Geometry* (1946) and his autobiography, *Souvenirs d'apprentissage* (1992, *The Apprenticeship of a Mathematician)*. The three volumes of his *Oeuvres scientifiques* (Collected Papers) were published in 1980.

WENDELIN WERNER
(B. SEPT. 23, 1968, COLOGNE, W.GER. [NOW GERMANY])

Wendelin Werner is a German-born French mathematician awarded a Fields Medal in 2006 "for his contributions to the development of stochastic Loewner evolution, the geometry of two-dimensional Brownian motion, and conformal theory."

Werner received a doctorate from the University of Paris VI (1993). He became a professor of mathematics at the University of Paris-Sud in Orsay in 1997 and part-time at the École Normale Supérieure in Paris in 2005.

Brownian motion is the best-understood mathematical model of diffusion and is applicable in a wide variety of cases, such as the seepage of water or pollutants through rock. It is often invoked in the study of phase transitions, such as the freezing or boiling of water, in which the system undergoes what are called critical phenomena and becomes random at any scale. In 1982 the American physicist Kenneth G. Wilson received a Nobel Prize for his investigations into a seemingly universal property of physical systems near critical points, expressed as a power law and determined by the qualitative nature of the system and not its microscopic properties. In the 1990s, Wilson's work was extended to the domain of conformal field theory, which relates to the string theory of fundamental particles. Rigorous theorems and geometrical insight, however, were lacking until the work of Werner and his collaborators gave the first picture of systems at and near their critical points. Werner also verified a 1982 conjecture by the Polish mathematician Benoit Mandelbrot that the boundary of a random walk in the plane (a model for the diffusion of a molecule in a gas) has a fractal dimension of 4/3 (between a one-dimensional line and a two-dimensional plane).

Werner also showed that there is a self-similarity property for these walks that derives from a property, only conjectural until his work, that various aspects of Brownian motion are conformally invariant. His other awards include a European Mathematical Society Prize (2000) and a Fermat Prize (2001).

SHING-TUNG YAU

(B. APRIL 4, 1949, SWATOW, CHINA)

Chinese-born mathematician Shing-Tung Yau was awarded the Fields Medal in 1983 for his work in differential geometry.

Yau received a Ph.D. from the University of California, Berkeley, in 1971. Between 1971 and 1987 he held appointments at a number of institutions, including Stanford (Calif.) University (1974–79), the Institute for Advanced Study, Princeton, N.J. (1979–84), and the University of California, San Diego (1984–87). In 1987 he became a professor at Harvard University, Cambridge, Mass.

Yau was awarded the Fields Medal at the International Congress of Mathematicians in Warsaw in 1983 for his work in global differential geometry and elliptic partial differential equations, particularly for solving such difficult problems as the Calabi conjecture of 1954 for both the Kähler and Einstein-Kähler metric cases, the positive mass conjecture (with Richard Schoen), and a problem of Hermann Minkowski's concerning the Dirichlet problem for the real Monge-Ampère equation. In the early 1980s Yau and William H. Meeks solved an open question remaining from Jesse Douglas' work on the Plateau problem in the 1930s.

Yau's publications include *Non-linear Analysis in Geometry* (1986) and, with Robert Greene, *Differential Geometry* (1993).

GLOSSARY

analytic geometry Study of geometry in terms of algebra and a coordinate system.

angle The space formed between two lines or rays diverging from a single vertex.

arc A section of a circle that includes some part of the circumference.

area The measure of a region enclosed on a two-dimensional surface (plane).

axiom A maxim broadly accepted as self-evident that serves as a basis for further analysis.

chord A line segment connecting two points on a curve.

circle A two-dimensional geometric curve in which all points are equidistant from one central point.

circumference The boundary containing a circle.

cone A solid with lines extending from every point on the edge of a planar base to a common vertex.

congruence Sameness in shape and size. Congruent figures can be superimposed if oriented in the same direction.

conic section A planar curve (i.e., an ellipse, hyperbola, or parabola) formed by the intersection of a plane with a right circular cone.

diameter A type of chord that runs through the centre of a figure. In a circle, the diameter runs through the central point and connects two points on the circumference.

differential geometry The study of curves and surfaces using algebra and calculus.

Euclidean geometry Study of planes and solids based on five axioms developed by Euclid.

fractal Complex and often irregular geometric shape composed of smaller parts that resemble the whole (i.e., a fractal is self-similar). A snowflake is an example of a naturally occurring fractal.

fundamental group A group of a topological space in which curves are combined by a geometric operation.

homeomorphism A continuous one-to-one mapping between two surfaces or figures.

isometry A map that preserves distance between points of one metric space in an image of those points.

metric space A set with a function specifying distance (metric) between two elements of the set.

parallelogram A polygon with four sides whose opposing sides are equal in length and parallel to each other.

plane A two-dimensional surface.

polygon A plane enclosed by three or more connected and uncrossed lines. Triangles, squares, and pentagons are all examples of polygons.

projective geometry The study of geometric figures and the images that result from projecting them onto another surface.

Pythagorean theorem Describes the relationship

between the three sides of a right triangle. The theorem states that the square of the length of the hypotenuse is equal to the sum of the squares of the lengths of the remaining two sides.

radius A line segment connecting a point on the circumference of a circle to the centre of the circle.

rectangle A parallelogram with four right angles.

similarity Having the same shape and angle measures but varying proportionally in size.

solid A three-dimensional figure. Cubes, spheres, and cylinders are examples of solids.

square A polygon consisting of four sides of equal length that are joined to form four 90° angles. A square is also an equilateral rectangle and an equilateral, equiangular parallelogram.

tangent A line or curve that touches a curve at a single point.

theorem A statement that must be proved.

topology The study of geometric properties that remain unchanged when one shape is elastically transformed into another without tearing or gluing any section.

triangle A three-sided polygon.

vector A line segment specifying direction and magnitude.

volume The measure of the size of a three-dimensional figure (solid).

Bibliography

General History

The best overview in English of the history of geometry and its applications consists of the relevant chapters of Morris Kline, *Mathematical Thought from Ancient to Modern Times* (1972, reissued in 3 vol., 1990), which can be supplemented, for further applications, by *Mathematics in Western Culture* (1953, reissued 1987). Three other useful books of large scope are Petr Beckmann, *A History of* π, 4th ed. (1977, reissued 1993); Julian Lowell Coolidge, *A History of Geometrical Methods* (1940, reissued 1963); and David Wells, *The Penguin Dictionary of Curious and Interesting Geometry* (1991). A fine survey at a college level of the various branches of geometry, with much historical material, is David A. Brannan, Matthew F. Esplen, and Jeremy J. Gray, *Geometry* (1999).

Ancient Greek Geometry

The standard English editions of the Greek geometers are those prepared by Thomas Little Heath beginning in the 1890s. They contain important historical and critical notes. Most exist in inexpensive reprints: *Apollonius of Perga: Treatise on Conic Sections* (1896, reissued 1961); *The Works of Archimedes* (1897, reissued 1953); *Aristarchus of Samos, The Ancient Copernicus* (1913, reprinted 1981); and *The Thirteen Books of Euclid's Elements*, 2nd ed., rev. with additions, 3 vol. (1926, reissued 1956). The historical material has been shortened and simplified, and its coverage extended, in *A History of Greek Mathematics*, 2 vol. (1921, reprinted 1993).

Further information about technical-historical points—for example, the lunules of Hippocrates—may be found in Wilbur Richard Knorr, *The Ancient Tradition of Geometric Problems* (1986, reissued 1993). The epistemology of Greek geometry can be approached via the editor's introduction to and the text of Proclus, A Commentary on the *First Book of Euclid's Elements,* trans. and ed. by Glenn R. Morrow (1970, reprinted 1992).

ANCIENT NON-GREEK GEOMETRY

Other ancient geometrical traditions are covered in A.K. Bag, *Mathematics in Ancient and Medieval India* (1979); Richard J. Gillings, *Mathematics in the Time of the Pharaohs* (1972, reprinted 1982); Joseph Needham, *Mathematics and the Sciences of the Heavens and the Earth* (1959), vol. 3 of *Science and Civilization in China*; and B.L. van der Waerden, *Science Awakening*, 4th ed., 2 vol. (1975).

GEOMETRY IN ISLAM

Aspects of the extensive development of geometry by Islamic mathematicians can be studied in J.L. Berggren, *Episodes in the Mathematics of Medieval Islam* (1986). Otherwise, the best route to a survey is through the relevant chapters in vol. 2 of Roshdi Roshed (Rushdi

Rashid) (ed.), *Histoire des Sciences Arabes*, 3 vol. (1997), and the articles on Arab mathematicians and astronomers in Charles Coulston Gillispie (ed.), *Dictionary of Scientific Biography*, 18 vol. (1970–90).

RENAISSANCE GEOMETRY AND APPLICATIONS

J.L. Heilbron, *Geometry Civilized: History, Culture, and Technique* (1998, reissued 2000), considers examples of geometry from some modern cultures as well as from the ancient Mediterranean and gives examples of the development of Greek geometry in the Middle Ages and Renaissance. A more advanced book along similar lines, but with more restricted coverage, is Alistair Macintosh Wilson, *The Infinite in the Finite* (1995). James Evans, *The History and Practice of Ancient Astronomy* (1998), is by far the best introduction to the theoretical and instrumental methods of the old astronomers. Albert van Helden, *Measuring the Universe* (1985), describes the methods of the Greeks and their development to the time of Edmond Halley. John P. Snyder, *Flattening the Earth: Two Thousand Years of Map Projections* (1993, reissued 1997), gives the neophyte cartographer a start. J.V. Field, *The Invention of Infinity: Mathematics and Art in the Renaissance* (1997), contains an elegant account, in both words and pictures, of the theory of projection of Brunelleschi, Alberti, and their followers.

GEOMETRY AND THE CALCULUS

The transformation of mathematics in the 17th century can be followed in Carl B. Boyer, *The Concepts of the Calculus: A Critical and Historical Discussion* (1939, reissued 1949; also published as *The History of the Calculus and Its Conceptual Development*, 1949, reissued 1959), largely superseded by Margaret E. Baron, *The Origins of the Infinitesimal Calculus* (1969, reprinted 1987); Michael S. Mahoney, *The Mathematical Career of Pierre de Fermat* (1601–65) (1973); and René Descartes, *Discourse on Method, Optics, Geometry, and Meteorology*, trans. by Paul J. Olscamp (1965, reissued 1976). This last work, which ranks among the most important books on natural philosophy and mathematics ever written, repays the effort required to master its idiom.

AXIOMATIC EUCLIDEAN AND NON-EUCLIDEAN GEOMETRY

Roberto Bonola, *Non-Euclidean Geometry*, 2nd rev. ed. (1938, reissued 1955), contains a thorough discussion of the work of Saccheri, Gauss, Bolyai, and Lobachevsky as well as a major text from each of the two founders of non-Euclidean geometry. David Hilbert, *Foundations of Geometry*, 2nd ed., trans. by Leo Unger and rev. and enlarged by Paul Bernays (1971, reissued 1992), is an excellent and accessible English translation.

INDEX

A

B

C

D